# STAR SETTLERS

# STAR
# SETTLERS

THE BILLIONAIRES, GENIUSES, AND CRAZED
VISIONARIES OUT TO CONQUER THE UNIVERSE

• •

## FRED NADIS

PEGASUS BOOKS
NEW YORK LONDON

STAR SETTLERS

Pegasus Books, Ltd.
148 West 37th Street, 13th FL
New York, NY 10018

Copyright © 2020 by Fred Nadis

First Pegasus Books hardcover edition August 2020

Interior design by Sabrina Plomitallo-González, Pegasus Books

ISBN: 978-1-64313-448-2

10 9 8 7 6 5 4 3 2 1

Printed in the United States of America
Distributed by Simon & Schuster

# CONTENTS

• • •

# THE STAR CHILD REBORN

• •

The geography of Santa Barbara, where I now live, is defined not only by the narrow strip of developed land between the coastal mountains and the Pacific Ocean—but also by the aerospace industry. One hundred miles to the south, at NASA's Jet Propulsion Laboratory (JPL), robotic space vehicles are designed and missions overseen, sixty miles to the north, SpaceX launches Falcon rockets at Vandenberg Air Force Base. After launches, in my neighborhood, rocket plumes angle in the sky over the mountains.

Ten miles up the coast in Goleta, in the 1960s at Delco Electronics, engineers designed and tested Lunar Rover Vehicles. The former Delco buildings now house other high-tech companies, also with defense industry ties. Nearby are Northrop Grumman's Astro Aerospace Manufacturing and Integration, ATK Aerospace Systems (satellite technologies), and Deployable Space Systems (solar array panels for spacecraft). Travel west 1.3 miles through the wetlands that border the airport, an area where archeologists still find Chumash Indian artifacts, and you are at the University of California.

On an April morning in 2018, while birds sang outside a conference room at the university—chirps blending with the back-up beeps of a maintenance vehicle—UCSB's Experimental Cosmology Group was meeting. With funding from NASA, this was one of the teams participating in Yuri Milner's "Breakthrough Starshot" with its goal of propelling minute, one-gram, spacecraft at about 20 percent the speed of light toward Alpha Centauri. The goal is to launch the "starchips" by 2040 on their approximately twenty-two year journey. The Cosmology Group's lead scientist, Dr. Philip Lubin, developed the concept for using laser arrays to propel the sailed starchips through interstellar space.

Bearded, short, broad-shouldered, energetic, Lubin worked a slide display for the group's weekly conference, choosing between folders with names such as: "Along the Path to Interstellar Flight," "Worms in Space 2017," and "The Meaning of Life." He opened a slide that plotted a wide range of organisms' "basal" (resting) and "torpid" (hibernating) metabolic rates versus mass and noted, wryly, that human passengers, for now, remained "a disaster for interstellar travel."

For practical reasons, the group has chosen hardier and smaller organisms as the best candidates for early interstellar travel. The favorites are the tardigrade (or tiny "water bear"—just visible in a petri dish) and a species of nematode, both of which can be dehydrated and later revived. Tardigrades, which can handle intense shifts in temperature and pressure as well as exposure to radiation, have lasted up to ten days unprotected in outer space, and can survive as long as a century in suspended animation. Should humans, ultimately, prove biologically unsuited to long space voyages (or to suspended animation), more imaginative speculations might salvage the dream. Lubin alluded to these speculations when he turned to his team's space theologian, Michael Waltemathe, based at Ruhr University in Germany, and joked, "We need a subgroup to redesign humans."

One of Yuri Milner's stated goals for "Breakthrough Starshot" is to determine if any of the planets in the Alpha Centauri triple-star system are habitable and perhaps already inhabited. The Starshot website asks, "When we find the nearest exo-Earth, should we send a probe? Do we try to make contact with advanced civilizations? Who decides? Individuals, institutions, corporations, or states? Or can we as species—as a planet—think together?" Ultimately, many are hoping that humanity or redesigned humans, i.e. transhumans, can make the journey. Interstellar flights are no longer entirely a fantasy. (Particularly if you are a nematode.)

Beyond the question "can we actually do it?" with which the UCSB Cosmology Group and other astrophysicists worldwide have begun to grapple, is the further question "why should we do it?" One of the reasons the cosmology group had a theologian in its midst was to contemplate the ethical issues the spaceflight revolution may raise— whether in a century, five hundred years, or a millennium. The goal of spacefaring, on one level, represents a simple desire for human survival, a feint to avoid destruction, but on another signifies enormous hubris: humans or human/mech offspring will seed the stars and offer immortality for the species. While space exploration and compiling knowledge about the cosmos seem like normal science at work, the added desire to "get out there" is often driven not by mere curiosity but a sense of destiny. As an idea, it can slowly grow on you.

As discussion in the UCSB conference room segued into the highly technical realm of laser specifications, I had time to wonder what had prompted me to be there—after all, I'd managed to pick up my humanities degrees in college with very few classes in math or the sciences. The most obvious connection I had with the gathered scientists was science fiction. Scratch any astrophysicist interested in interstellar flight and you're likely to find a "hard science fiction" fan. For example, in the UCSB conference room was Andrew Higgins, a visiting physicist from McGill University, who commented that he was taking an "interstellar sabbatical" at UCSB because as an adolescent he had read physicist Robert L. Forward's *Flight of the Dragonfly* (1984), a science fiction book which offered a schematic for laser light and sail-driven starships.

Unlike these scientists and many of my peers, I was a latecomer to the game. In my kid years, I was more of a comic book fan—if you handed me a copy of *Richie Rich*, *The Flash*, *Dottie Dot*, *Tales from the Crypt*, or *Archie and the Gang*, I'd read it—yet I was slow to warm to science fiction. In seventh grade, for example, I had friends who had formed the "Rutabaga Club," who liked to chant "rutabaga, rutabaga, rutabaga," at basketball games. One member, Feinstein, whose dad was a mathematician, beat me in chess using "fool's mate." Then there was Perlman. I was invited to his house one afternoon, and I examined the science fiction paperbacks on his shelves. It was all that the Rutabaga Club members read. Stuff like Roger Zelazny's *Lord of Light*, John Brunner's *Stand on Zanzibar*, and Arthur C. Clarke's *Childhood's End*. I liked the psychedelic covers, but didn't get the attraction. A mutual decision was reached: I didn't join the Rutabaga Club.

But science fiction creeps up on you—or more exactly crept up on culture at large like a cyber cowboy hacking the electronic protection on corporate ice-laden gates. Every punk rocker friend of mine in 1980s Lower East Side Manhattan worth their weight in studded S&M regalia, for example, had a copy of J. G. Ballard's *Crash* in their walkup. Although I never finished *Crash*, I came to appreciate Ballard's other novels, such as *The Drowned World* (1962), in which London, thanks to global warming, has been claimed by waters and jungles and transformed into a metropolis from a lost world epic. Slowly, I became hooked on the genre.

While science fiction mixes adventure with science, at its base lurks a mythic idea: the death of old worlds and birth of the new, i.e., apocalypse.[1] For example, even though the Rutabaga Club members and I were only eleven when it premiered, we saw Stanley Kubrick's 1968 film *2001: A Space Odyssey*. Toward its end, weary astronaut Dave Bowman, on a mission to study an alien obelisk, passes through a psychedelic "stargate." Bowman, in his red spacesuit, then enters a chamber with glowing floor, sumptuous furnishings, and statues of nymphs. He sees an older version of himself, in a dark robe, stoically dining. As the old astronaut draws his last breaths in a wide bed, the obelisk appears, and he then transforms into an embryo, hovering in space near an alluring vision of (Mother) Earth. To the fanfare of Richard Strauss's "Also Sprach Zarathustra," the Star Child is born.

As in *2001*, science fiction writers often render the end of the world and the birth of the new. The end is sudden (an asteroid crash, nuclear conflagration, or alien invasion) or it involves a slow death rattle, usually from cascading biomedical and environmental disasters. And so, to my long explanation of why I was in a conference room where the lead scientist had a computer folder labeled "Worms in Space"—I can add one more suspect: the apocalyptic myth lurking in the DNA of science fiction.

In our dystopian present, dwelling on impending disaster appears less like grand adolescent fantasy and more like pragmatism. Planning to leave the Earth coincides neatly with worries about an approaching doomsday. Decades ago, astrobiologist Carl Sagan offered the epigram, "All civilizations become either spacefaring or extinct." In 2007, Yale astrophysicist J. Richard Gott estimated that we had forty-five years to establish a Mars colony if we wanted to ensure species survival, based on the assumption that the then forty-five-year-old space program would continue at least that long in the future.[2] Scientists Freeman Dyson and Stephen Hawking also supported the idea of an off-Earth refuge. In 2016, industrialist Elon Musk announced our species must become "multi-planetary" to evade extinction—presumably from an asteroid strike (AKA "the big one"), or nuclear or ecological disaster.

While the powerful subculture of scientists, engineers, industrialists, and intellectuals nurturing the spacefaring dream may prove to be a true vanguard, their notions have a familiar ring. Over a century ago, Russian cosmists such as Nikolai Fedorov and Konstantin Tsiolkovsky declared the Earth was not the natural home for humans. No, humanity's destiny was to break free of gravity, and evolve into a godlike space-dwelling race. Fedorov, steeped in Russian Orthodox mysticism, also thought the true goal of mankind was to defeat death by resurrecting all the dead. We could then become immortal and abandon bodies for new forms. This was, needless to say, grand apocalyptic thinking.

As of this writing, the Russian cosmists' twinned obsessions with immortality and spaceflight are common among Silicon Valley billionaires, some of whom are preparing underground bunkers as well as a future in which the wealthy can upload their consciousness into supercomputers to live forever. For a growing number of entrepreneurs, funding a rocket program also has become de rigueur. Companies such as Blue Origin owned by Jeff Bezos—the world's wealthiest person as of 2019, and Elon Musk's SpaceX, might help realize some of science fiction's wilder fantasies. Musk claims that all his commercial enterprises have been created to fulfill his dream of colonizing Mars. In his vision, within a hundred years, a fleet of rockets will make trips every two years to establish a colony with a population from 80,000 to one million people (depending on the speech) on the Red Planet. Musk has quipped, "I'd like to die on Mars, just not on impact."

A decade ago, author Tom Wolfe scolded NASA for relying on engineers as the dry voice of the Mercury, Gemini, and Apollo programs and failing to draft "philosophers" who could unleash "The power of the Word."[3] The myth was buried in space dust. Excitement over the Apollo moon program clearly was on the wane by 1972 when Apollo 14 Commander Alan Shepard, holding a smuggled golf club, teed off on the Moon and bantered with ground control in Houston about whether he had sliced his shots. As the Vietnam War dragged on, the U.S. industrial base dwindled, the Bee Gees became popular, environmental concerns mounted, and the Space Shuttle Columbia exploded, interest in space exploration fizzled. In 1989 George H. W. Bush's proposal to send a mission to the Moon and then Mars elicited a yawn; his son George W. Bush received a similar ho hum response to his Moon to Mars plan.

With the advent of environmentalism, as the "Small Is Beautiful" mindset stole center stage and solar power, Birkenstock sandals, and bicycles became the answer to any question—when even Mr. Spock sought to save the whales—space enthusiasts struck back. Did we want to retreat into a world of "no growth" and stagnation or pursue a dynamic course? Former Nazi aerospace engineer Krafft A. Ehricke insisted in an influential 1971 essay that humanity was designed to leave the Earth's "closed system," which couldn't handle industrial by-products, and to follow "the Extraterrestrial Imperative" into a cosmos of limitless resources. For those who cared, we could even beam solar power back to the Earth!

Ehricke believed that the Extraterrestrial Imperative, the need for humans to exit planet Earth, was part of a greater plan. His idea, similar to that of the cosmists, slowly caught on. We could solve Earth's environmental problems—by leaving. At a 2009 NASA sponsored conference on "Cosmos and Culture" space enthusiast Howard Bloom urged, "Evolution is shouting a message at us. Yes, evolution herself. That imperative? Get your ass off this planet. Get your asses, your burros, your donkeys, and as many of your fellow species as you can."[4] Less stridently, in 2019 Jeff Bezos argued that the Earth must eventually be zoned "light industrial" while trillions of humans move onto orbiting space colonies throughout the solar system. From this vast population, thousands of geniuses like Einstein and Mozart might emerge.

Yet though boundlessly optimistic and idealistic, the spacefaring narrative has its shadow. Requiring immense resources, the conquest of space has always been highly militarized. The tangle of aerospace and military contractors on my short stretch of Highway 101 on the West Coast illustrates this connection—almost as clearly as the U.S. Congress's 2019 decision to establish the Space Force as a new branch of the military. Likewise, the UCSB Experimental Cosmology Group, in attempting to bring multiple lasers into phase for spaceship propulsion, is building on technology related to weaponry experiments at DARPA—the Defense Advanced Research Projects Agency. And for much of its earlier successes, the U.S. space program had depended on Wernher von Braun and his team of Nazi V-2 rocket experts (including Ehricke) that were brought to the U.S. after World War Two. In addition to the thousands of civilians the V-2 rockets killed in Britain and in Belgium during the war, as many as 20,000 enslaved workers died at the production site, complete with its own concentration camp, Mittelbrau-Dora. No price was too high for the license to design rockets that one day would reach the heavens.

Intergalactic warfare has always been popular in science fiction, but it is rendered several dozen steps removed from current reality, the result tending toward escapist entertainment. Taking to the stars, whether in swashbuckling or deeply philosophic fashion, is a science fiction mainstay. Some leading writers, however, have their doubts.

Kim Stanley Robinson, who provided a blueprint for terraforming Mars in his 1990s Mars trilogy, argued in the novel *Aurora* (2015) that humanity was simply not meant for distant spacefaring. In *Aurora*, a multigenerational starship arrives at its destined exoplanet, and, as a plague spreads, the would-be colonizers realize that it is a world with a fundamentally hostile biome. Finding colonization impossible, the settlers do the unthinkable, and return to Earth. If Robinson is correct, we may be consigned, for better or worse, to life on "Spaceship Earth" and its immediate neighborhood.

Yet spacefaring remains a powerful dream. What other idea could unite Russian mystics in the last days of the tsars, Nazi engineers, Jewish immigrant teenagers such as Isaac Asimov in 1930s Queens, 1970s techno-hippies with groovy ideas about space colonies, Afrofuturist authors such as Octavia Butler, and the new crop of billionaires relying on their own fortunes to pursue manned spaceflight and help fund programs like the one I was visiting at UCSB? What once was fringe thought—escaping to the stars—has been inching toward the center. A potentially profound cultural change appears to be underway, as we shift from thinking of ourselves as an earthbound species to one of (potential) spacefarers. Whether we are worthy candidates for dispersal through the solar system or galaxy remains an open question.

# CHAPTER ONE

# MARS MANIA 1.0

• • •

"I am of another world," I answered,
"the great planet Earth, which revolves around our common sun and
next within the orbit of your Barsoom, which we know as Mars."
—EDGAR RICE BURROUGHS, *A Princess of Mars*, 1911

"Mars is fine but it's a fixer-upper planet."
—GWYNNE SHOTWELL, president of SpaceX, 2018

At a 2018 debate at the SETI Institute in Menlo Park, California, Robert Zubrin, a brash man of rumpled appearance, paced the stage; his hands became agitated as he explained the costliness and downright craziness of NASA's Interplanetary Protection Program. This policy began in 1967, when the United States signed the Outer Space Treaty which stipulated that space exploration not lead to contamination of another planet ("forward contamination") or of the Earth ("back contamination"). Committed space settlement advocates do not favor this cautious approach. Zubrin, an aerospace engineer and the founder of the Mars Society, insisted that the treaty needlessly added costs to space missions and stifled exploration. For billions of years, he noted, our planet has been bombarded with meteors from Mars—bacteria, if it exists or existed on Mars has been "coming in flocks" all along so there's no need to fret over microbes smuggled out as hitchhikers on our equipment or as part of soil sample return missions.

In addition to this concern about "back contamination," "forward contamination," Zubrin argued, was also not that serious. In fact, we didn't really have to worry about the transfer of life between planets. Should we bring bacteria to Mars, Zubrin would rather call it "fertilizing Mars" or "enlivening Mars," not contaminating it. Zubrin's opponent, John Rummel, a former Planetary Protection Officer at NASA, unsurprisingly, differed.

During this debate in April 2018, the Curiosity rover, or Mars Science Laboratory, NASA's third on the Red Planet, was continuing its $2.5 billion mission: to search for evidence of microbial life on Mars—present or past—and to survey surface conditions to prepare for human landings. It was also beaming back majestic panoramas of the planet, reminiscent of America's dry west, making this oddly familiar neighboring planet seem ideally cast as a new frontier. Mars soil sample return missions also were being planned. (The proposed 2020 NASA budget included missions to bring Mars soil samples to Earth by 2031.)

No one at the Jet Propulsion Laboratory in charge of the Mars rovers expected to find intelligent life or monstrous beings in Martian craters. But such fantastic dreams only slowly fade. In 1976, when Viking 1 snapped photographs of Mars from orbit, NASA technicians noted that one rock formation, coincidentally, resembled a human face. The notion that the "Face on Mars" was evidence of a Martian civilization became a long-running theory. Tabloid journalists and conspiracy theorists insisted that other geometric patterns on the planet's surface also indicated the work of intelligent beings. The *National Enquirer* and allied websites have noted dozens of them. (While such sightings have slackened over the decades, 2018 was a bumper year. To take one example, on April 30, 2018, my Yahoo news feed offered another face on Mars—this one of a "Warrior Woman Statue.")

Fittingly, it was a similar set of optical illusions, gathered by scientists in the late Victorian era—and more specifically, one map, the 1877 map by Italian scientist Giovanni Schiaparelli that first encouraged the appetite for all things Martian—and indirectly inspired the Mars Science Laboratory rover now leaving wheel tracks in the red dust of Mars—not to mention the blooming of the citizen-led Mars

Society with its 10,000 members planet-wide, or those itching to reserve one of the hundred seats on SpaceX's planned Mars "Starship" transporters when colonization begins.

The best nineteenth century viewing of Mars came during the 1877 opposition. ("Oppositions," when the Earth on its orbit nears Mars on its wider orbit, occur approximately every two years and two months.) That year, an astronomer in Washington, DC, discovered that Mars had two moons, and in Milan, astronomer Giovanni Schiaparelli was startled to see the planet crisscrossed with canali, or channels. Their geometric rigor implied, to Schiaparelli, a civil engineer turned astronomer, that there might be an intelligent species on the planet that had planned and excavated the channels. He carefully mapped them, and left it for others to explain what they were. Schiaparelli was cautious, but in defense of his discovery wrote to a skeptical colleague, "It is as impossible to doubt their existence, as that of the Rhine on the surface of the Earth."[1] Although Schiaparelli's observations were controversial, they were not rejected outright; indeed, other astronomers also began to report seeing canals. Schiaparelli's map of the surface of Mars with its channels gained popularity.

Not everyone believed in the canali. During the same opposition of 1877, astronomer Nathaniel Green trained a telescope on Mars and drew a map completely different from Schiaparelli's—it included swirling nuanced colors, few identifiable features—and no canals. Green politely wrote letters to Schiaparelli suggesting he might be in error, but the Italian astronomer held firm. In the decades that followed, astronomers continued to debate the merits of the maps of Schiaparelli and Green.

The dispute about the canals proceeded calmly for several decades—and then became quite heated. While the existence of channels was not outrageous, some found infuriating the further argument that they indicated Mars had nurtured intelligent life—indeed a society capable of launching major hydraulic projects to combat a drying climate. Schiaparelli, regarded as a keen-eyed observer, had a formidable group of allies and popularizers. One of his most influential backers in the scientific community was Camille Flammarion, an astronomer, balloonist, science popularizer, and firm believer in extraterrestrial life.

Flammarion was born in a small village in France in 1842 and as a child developed his fascination with astronomy—he recalled borrowing a pair of opera glasses at age eleven, and seeing "Mountains in the moon as on the earth! And seas! And countries! Perchance also inhabitants!" As a young man he became an apprentice to an engraver in Paris, while continuing his studies in night school. A physician treating him for exhaustion discovered the teenager's lengthy manuscript, "Cosmogonie Universelle," about the "origin of the world," and with his recommendation, Flammarion, at age sixteen became a pupil astronomer (doing computations) at the Paris Observatory. His career as an astronomer, meteorologist, popular author, and cosmic visionary quickly followed. He came to prominence at age twenty with the tract, *The Plurality of Inhabited Worlds* (1862), which hit a nerve with his argument for the likelihood of life on other planets. In 1864, he added *Worlds Imaginary and Real* in which he reviewed the history of ideas about extraterrestrial life. Clearly, he was certain they were out there—and in this, he was then (and likely now) in the scientific mainstream.

For centuries, leading thinkers reasoned that with innumerable planets orbiting innumerable stars, the "purpose" of these other planets, like our own, was to harbor life. In 1600, Giordano Bruno was burned at the stake for asserting this and other heresies. Isaac Newton, Gottfried Wilhelm Leibniz, Immanuel Kant, Voltaire, Benjamin Franklin, Edmond Halley (the comet discoverer), and astronomer William Herschel and his offspring were among the supporters of "pluralism," the belief that there were many inhabited worlds. Seventeenth-century French philosopher Bernard le Bovier de la Fontenelle had advanced this case in his *Conversations on the Plurality of Worlds* (1686).

Relying on advancing scientific knowledge and an ample imagination, Flammarion added flesh to such speculations. In one of his catalogs of non-earthly beings, Flammarion populated Delta Andromedae—a planet with an atmosphere heavier than our air but lighter than water—with rose-colored, floating citizens who survived breathing its nutritious air with overworked lungs; he also offered, on the planet Orion, with its seven suns, plant-like men that moved on starfish feet, and another organism, resembling a chandelier, that could break into pieces and vanish only to later reassemble. Flammarion insisted that otherworldly organisms might have ultraviolet or infrared eyes, an "electric" sense, and spectroscopic abilities.[2] He offered such wonders in a tone that mixed matter-of-fact detail with poetry. Flammarion's *Plurality of Inhabited Worlds* went through thirty-three editions by 1880 and remained in print until 1921.

When Flammarion came of age, in nineteenth-century France, engineers and scientists were culture heroes. French philosopher Henri de Saint-Simon had earlier noted, "A scientist, my dear friends, is a man who foresees; it is because science provides the means to predict that it is useful, and the scientists are superior to all other men." Saint-Simone's disciple Auguste Comte, who sought to free Enlightenment thought and its benefits from the grasp of elites, was even more emphatic about science and engineering as the ultimate forces for good in society.

Newspapers and magazines were flooded with science articles and reports from the Académie, lecturers abounded, and publishing companies such as Larousse, Hachette, and Flammarion (Camille Flammarion's brother Ernest was the founder) offered scientific fare, while Jules Verne, who at times referred to Flammarion in his novels, helped forge a new genre that intermixed science with semifantastic adventure. Another French science popularizer of the time, Louis Figuier wrote in 1867, "Science is a sun: everybody must move closer to it for warmth and enlightenment."[3] Like Comte and Figuier, Flammarion treated science as a redemptive force to be widely dispersed. He was certain that the more that astronomy progressed the greater its appeal. "It leaves the realm of figures and comes alive. Filled with wonder, we see the spectacle in the skies transfigured . . . The science of the stars ceases being the secret confidant of a small number of experts; it penetrates everyone's mind, it illuminates nature."[4]

By age twenty-three, while continuing to work in observatories, Flammarion became an avid balloonist, then president of the French Aerostatic Society, and he made numerous ascensions while conducting meteorological experiments. His balloon outings included his honeymoon, a flight in 1874 with his new wife Sylvie Petiaux-Hugo Flammarion (a grand-niece of Victor Hugo), and culminated in a book about the Earth's atmosphere. In 1877 Flammarion published his *Catalog of Double Stars* that became a valuable tool for astronomers, and followed this with the lavishly illustrated *Popular Astronomy* (1880)— which sold over one hundred thousand copies. In it, he argued there was likely life not only on Mars but also the Moon. In 1877, Flammarion founded the French Society of Astronomy and in 1882 the magazine *L'Astronomie*. He also organized France's first observatory open to the public in Paris's Latin Quarter. Central to the science of astronomy in the late nineteenth century, Flammarion was forgiven his outsized speculations. In fact they were required: reporters flocked daily to his apartment in Paris (its ceiling decorated with the signs of the zodiac) to gather quotes on matters small and large.

Unlike Robert Zubrin and twenty-first century members of the Mars Society, Flammarion was less interested in depicting a future where humans could settle other planets than in stressing that the universe was full of wonders—including the news that humans and many other beings already populated the planets. He was intensely interested in the new field of psychic research, held séances in his parlor, and alternated books on popular astronomy with books on psychic phenomena as well as fiction that fused both realms of speculation. Flammarion's cosmology—which sought to blend the physical and spiritual universes—in addition to unveiling the "nuts and bolts" of the cosmos, offered a view in which the planets of this solar system and other star systems might be "heavens" or new Earths that humans went to after death where they were reincarnated in slightly more spiritual form.

He offered this vision in *Lumen* (1872), and later in *Uranie* (1891), in which the muse of astronomy takes Flammarion on a tour of the universe, and Flammarion recounts meeting on Mars a friend who had died young in a ballooning accident only to be reborn on the Red Planet. No spaceship was involved in these dream journeys, but rather a form of soul travel to other planets. With his ballooning adventures, séances, impassioned writing, public lectures, ample energy, large furrowed brow, and handsome looks, Flammarion gained wealthy admirers—many, but not all, female. In 1882, one such admirer, Monsieur Meret of Bordeaux gave Flammarion a mansion (including a stable with horses) that Flammarion converted into an observatory in Juvisy-sur-Orge to the southeast of Paris. It was inaugurated in 1887 with the emperor of Brazil in attendance.

It wasn't until the opening of his Juvisy observatory that Flammarion fixed his interest on Mars and added his influence to the debate over Mars's canals. In 1892, he published *La Planète Mars et Ses Conditions D'habitabilité*, a survey of all the previous scientific studies of the planet of the past two centuries. The book marshaled all the available knowledge about Mars, including the planet's two moons, thin atmosphere, the waxing and waning of the planet's polar ice caps, its varying tilt on its axis, the seasonal color changes that suggested vegetation covered some of the planet's surface (an idea that originated with Flammarion and maintained credence well into the twentieth century), its length of day (24 hours, 37 minutes) that closely paralleled that of the Earth, and a review of the nearly 400 drawings of Mars that had been completed over the previous two centuries.

Flammarion noted that these drawings, which widely varied, relied on art as much as science. "It is extremely difficult to make a faithful drawing of what can be seen, because the forms are nearly always indefinite, diffuse, vague, without sharp outlines, and sometimes quite uncertain."[5] Flammarion admitted that efforts to clearly see the details of the planet through two atmospheres with even the best telescopes resulted in ambiguity. He affirmed that Mars included "streams" and "seas," but suggested, reasonably, that its age, greater than that of the Earth, explained its relative lack of water and great deserts. While Flammarion could only confirm one of the numerous canals that Schiaparelli and disciples had mapped, he hedged his bets. Schiaparelli was a much-respected figure, known for his keen eye, and Flammarion tried to imagine how such canali might exist. He also confronted Schiaparelli's report that the canali might appear, disappear, and, as mysteriously, sometimes "double." Could water, somehow, exist in a different chemical state to explain this behavior? Ultimately, Flammarion was swayed by a preference for observations that supported plurality. Suspecting the canali were not purely natural formations, he wrote, "the actual habitation of Mars by a race superior to our own is in our opinion very probable."

Flammarion's description of Mars gained Schiaparelli his most ardent champion: Percival Lowell. On Christmas day, 1892, a relative gave Lowell a copy of Flammarion's *La Planéte Mars*. The book inspired the wealthy Lowell, a Harvard graduate, and helped him fix on a vocation. Lowell had turned down a job teaching math at Harvard to travel through Asia. During his travels, he had served as a secretary to a diplomatic envoy, and penned dispatches and books on Korean and Japanese culture—including a work that examined occult beliefs and religious practices in Japan. When Lowell read Flammarion's *La Planéte Mars*, he found a new, fixed goal: to set up his own observatory and prove that life existed on Mars. In fact, Lowell wrote "Hurry!" on his copy of Flammarion's work, aware that the next opposition of Mars was 1894.[6]

Within two years, the Lowell Observatory, established with the help of astronomers and technicians on loan from Harvard, began operating outside Flagstaff, Arizona, and Lowell quickly published his own book, *Mars* (1895), which confirmed Schiaparelli and went further. Schiaparelli, he insisted, was not deluded—there was an elaborate network of irrigation canals on Mars, and these were clearly the work of a doomed intelligent species, led by brilliant engineers, trying to prolong life on a dying planet.

Like Flammarion, Lowell was a charismatic figure. At Harvard he had been the protégé of mathematician Benjamin Peirce—and he was a polished writer and speaker. Many journalists commented that they felt in the presence of greatness when they interviewed Lowell or heard him lecture. With his noble bearing, fine moustache, and stylish dinner jacket, he created a powerful effect, and his insights and witticisms won over his listeners. Unlike Flammarion, Lowell was not drawn to occult interpretations of extraterrestrial life (soon spiritualists would be channeling messages from Martians). Instead, Lowell's argument appealed to a modern secular sensibility. As a Darwinian, Lowell was certain that given appropriate conditions life would evolve anywhere in the universe. Lowell had no use for theological props, framing his theory within Pierre-Simon Laplace's "nebular hypothesis" of planetary formation, which proposed a primordial star-like gas cloud, rotating, condensing, and casting off matter that would eventually cool and harden as planets.[7] Older, smaller, and farther from the Sun, Mars would have cooled earlier than the Earth, speeding its desertification, and the loss of its atmosphere, so explaining the desperate plight of the highly intelligent and evolved Martians.

Lowell had his defenders among scientists, but many established astronomers were skeptical or outright hostile—preferring Schiaparelli's cautious position that while the canals appeared to be real, interpretations were open. Lowell, however, was unwavering in his theory that the canali offered definitive proof of life on Mars—any other explanation was cowardly and foolish. For several decades, Lowell's views held sway in the public debate waged in newspaper supplements and magazines. To Schiaparelli's map Lowell added his own globe of Mars, which included an even more elaborate and precisely geometric network of canals. In addition to interviews, and lectures, Lowell wrote two more books, *Mars and its Canals* (1906), dedicated to Schiaparelli, and *Mars as the Abode of Life* (1908). He never flagged in his assertions, even when many other astronomers insisted there were no canals to be seen—and one leading astronomer could remark that Lowell had become "a trial to sane astronomers."[8]

Lowell argued that the Martians, adapting to a steadily drying planet, had excavated the canals. As the planet's polar caps melted seasonally, the canals were opened to cultivate vast tracts of land, and it was these wide agricultural belts that appeared to observers as dark lines on the planet's surface. Lowell gave gripping lectures that described this heroic effort of a doomed civilization, and he insisted that any position but his was folly. Lowell enlisted influential friends such as sociologist Lester Frank Ward to bolster his theories about Martian civilization. He also relied on his family connections to give Lowell Institute talks and promoted his views in leading magazines such as the *Atlantic Monthly* and *Century*.

The dispute was long-running. When *Popular Astronomy* hailed Lowell as the era's leading astronomer, his establishment rivals were furious. Despite the anger on both sides, there was room for humor. Offering support for plurality, in 1894 Flammarion penned a letter that he claimed had been found in a meteorite that had crashed through the roof of the French astronomical society. Written from the point of view of a Martian scientist, the paper was titled, "Can Organic Life Exist in the Solar System Anywhere but on the Planet Mars?"[9] Weighing all the scientific evidence for life on Earth, the writer concludes that conditions, sadly, made the case for life on planets other than Mars impossible.

While Lowell developed his globe of Mars and its canals, skepticism over his account mounted, particularly as he taunted and outmaneuvered the leaders of large observatories including Edward C. Pickering at Harvard and William W. Campbell at Lick Observatory. Arguments erupted over the possibility that optical illusions had led Schiaparelli, Lowell, and their allies to interpret inchoate data as lines or canals. This theory, that the canals were misinterpretations of fuzzy data, gained strength in 1903, when E. W. Maunder and J. E. Evans ran an experiment with British schoolboys. Seated at different distances from a small drawing of Mars (about six inches wide) that lacked clear details, they were asked to reproduce the image. Most connected splotches into sharp lines. Of their experiment Maunder and Evans argued that the Martian "canals . . . have no more objective existence than those" drawn by the schoolboys.[10]

Lowell's response was scornful. He cited an experiment that Flammarion carried out with French schoolboys that created the opposite result, as well as his own tests of a telescope viewer's ability to discern penned lines at a distance proportional to views of markings on Mars. Lowell concluded that the British schoolboys had been misdirected and were following suggestions to ensure hoped-for results.

Maunder and Evans's experiment did not entirely destroy Lowell's credibility. In 1905, Lowell financed his assistant C. O. Lampland's expedition to the Andes to photograph Mars, under arduous conditions, through the thin atmosphere on Andean mountain peaks. When shown in lectures and published side by side with Lowell's drawings, these photographs made a strong case that the canali had a basis in the planet's features. Many astronomers and academics were convinced, as were journal editors. *Scientific American*, for example, ran an article titled "Canals of Mars Photographed," in June 1905. After this wave of success, Lowell published his second book, marshalling further evidence that a doomed race had engineered the surface of Mars.

Opposition continued. One of Lowell's most fervent opponents was Alfred Russel Wallace, codeveloper with Charles Darwin of the theory of evolution through natural selection. In 1907, Wallace published *Is Mars Habitable?* Although a Spiritualist who had defended as genuine forged photographs of fairies dancing with English schoolgirls, Wallace was also a clear-headed scientist. He also was convinced, a priori, that humanity was unique in the cosmos. Intelligent life had not evolved elsewhere. Wallace argued that if the Martian canals were natural features, they were the result of geological cracking, perhaps from volcanic activity. He argued further that there was not enough water stored in Mars's polar ice caps to irrigate the canals Lowell had mapped. Nor any conceivable way that the water, if so directed, could return to the poles seasonally. Likewise the planet was too cold and its atmosphere too thin to support life.

Two years later, Lowell's opponents firmly gained the upper hand. During the 1909 opposition with Mars, astronomers armed with powerful telescopes detected no canals—and greater details where canals earlier had been mapped. One of the chief observers was Eugene M. Antoniadi, a former assistant to Flammarion, and a previous believer in the canali. His "defection" gave his observations greater weight. During this 1909 opposition he worked with a 33-inch refractor telescope at Meudon Observatory (presumably on very clear nights in Paris) and concluded, after studying three dozen putative canals, that they corresponded to "real" features but not to canals.

Antoniadi's observations also were backed by George Ellery Hale at the Mt. Wilson Observatory in the United States. The canals, these observers insisted, were actually dark "irregular" features on the planet. Cartographers such as Nathaniel Green, who had rejected canals and insisted on swirling, irregular surfaces gained the high ground. The controversy, among scientists, largely was put to rest. Yet by then, the idea of the canals had become a staple of newspapers and early science fiction.

While Lowell pretended to be impervious to his peers' conclusions, he began to widen the research program at his observatory. As of 1905, Lowell began a search for a ninth planet, beyond Neptune, which he termed "Planet X." (Clyde W. Tombaugh finally identified Pluto in 1930 at the Lowell Observatory almost two decades after Lowell's death in 1916.) With his studies of Mars and search for "Planet X," Lowell saw himself on a holy quest—he had retreated to the desert for a clear vision, and as noted on his tomb, he remained a seeker to the end, "To see in the beyond requires purity . . . and the securing it makes him perforce a hermit from his kind . . . He will abandon cities and forego plains . . . Only in places raised above and aloof from men can he profitably pursue his search."[11]

Geographer Kristina Maria Doyle Lane has argued that the canal controversy was not simply a case of "bad science" versus "good science." In the late nineteenth century, telescopes brought Mars to the outer limits of observation. No one saw more than one or two "canals" on a given evening. The map of crisscrossed canali slowly evolved based on multiple sightings (just as had Green's map of swirling colors and vague shapes). Lane argued that in this context, astronomers who could add a new canal, that is, offer specific data, gained attention, whereas those who saw no distinct markings fell into the background.

The ensuing creation of canal-laden maps and globes of Mars added support to Schiaparelli's speculations.[12]

Whether it was good or bad science, Schiaparelli's "discovery" and Lowell's crusading unleashed a Mars mania at the turn of the century. This included a deluge of speculative fiction of all sorts about the planet. Mars became fodder for utopian thinkers, socialists, Spiritualists, and helped usher in science fiction as a unique genre. The basic premise came from Lowell by way of Schiaparelli: Mars was an aged, doomed world, peopled by brilliant mathematicians and engineers, who had mounted a massive project to irrigate their world. Many rather unmemorable novels and plays were written in the 1880s and 1890s featuring journeys to Mars and encounters with its human-like inhabitants and their carefully planned society. *A Message from Mars*, a hit play first mounted during the 1899 Christmas season featured a stingy miser converted to good deeds after a visit from an otherworldly Martian.[13]

Mars mania led to the sheet music for the 1901 hit "A Signal from Mars" (a two-step). In 1905, visitors to Coney Island's Luna Park could join a long line outside the Aerial Navigation Company for a simulated "Trip to Mars by Aeroplane." Martian literary fantasies varied. Alice Ilgenfritz Jones and Ella Merchant wrote *Unveiling a Parallel* (1893), a feminist utopia set on Mars, while Alexander Bogdanov's Bolshevik-inflected science fiction novel *Red Star* (1908) featured a Russian cosmonaut who discovers a perfected socialist society on Mars. In 1898, German author Kurd Lasswitz published *Two Planets*, describing a Martian excursion to Earth, with earthlings later escorted on a tour of Mars and its canal-centric civilization.

Of the early Mars books, H. G. Wells's *The War of the Worlds* remains the most influential. Wells was born in 1866, when Flammarion had already published his first book. Wells's fierce intelligence helped him, though born to the working class, to avoid a life of drudgery. His apprenticeship to a draper, at age thirteen, required him to work thirteen-hour days, with little time for reading or thought. The stifled genius hated the trade; his employer dismissed him for being "untidy and troublesome."[14] Wells's spotty education was sharpened, particularly in the sciences, when admitted to a teacher preparation program at London's Normal School of Science. T. H. Huxley (known as "Darwin's bulldog"), a faculty member, dazzled the young science student. After a bout of tuberculosis nearly killed him, Wells turned to writing. He applied his fierce Darwinist reasoning, imagination, and sensibility to his earliest—and most famous—novels, a quartet of scientific romances, *The Time Machine* (1895), *The Island of Dr. Moreau* (1896), *The Invisible Man* (1897), and *The War of the Worlds* (1898).

While much of Jules Verne's works, aimed at children, celebrated scientific advance, Wells's tales offered sharp social criticism and examined the double-edged nature of scientific knowledge. Wells's first novel, *The Time Machine* depicted a far future in which the classes had become separate species and also offered readers an imaginary voyage to the world's final heat death (a tour that Flammarion had experimented with in his earlier quasi-novel, *Omega*). Unlike the two books that followed, *The Island of Dr. Moreau* and *The Invisible Man*, *The War of the Worlds* featured no mad scientists, but instead the invasion of Earth by Martians desperate to escape their dying planet.

Wells's Martians did not resemble humans—or the enlightened souls with which Flammarion peopled the stars. Instead, they resembled highly intelligent octopi, sporting two clutches of tentacles beneath a giant head—making them the ancestor for all the Bug-Eyed Monsters (or BEMs) to eventually appear in pulp science fiction. Although clumsy in the Earth's stronger gravitational pull, these cephalopods from outer space were utterly indifferent to the suffering of the humans that they conquered with superior weaponry and intelligence. Theirs was an ecological conquest as well, as the Martian brought along at least one species of reddish plant (à la Flammarion) that began to overrun the English landscape. Of the Martians' destructive march through the English countryside and London, Wells remarked to a friend, "I'm doing the dearest little serial for Pearson's new magazine, in which I completely wreck and destroy Woking—killing my neighbours in painful and eccentric ways—then proceed via Kingston and Richmond to London, which I sack, selecting South Kensington for feats of peculiar atrocity."[15]

The book was chilling, believable, mocked imperialism, and asked readers to consider the arbitrariness underlying humankind's assumption of superiority and dominion over other species. It also critiqued the narrative tradition that depicted extraterrestrials as benign, angel-like overseers of lesser races. (The dialectic between treating extraterrestrials as space saviors or monsters has continued through the decades.) Wells shared Lowell's coldly materialistic view of the cosmos. Humanity, in Wells's vision, was responsible for its own destiny. The book was hugely popular. Pirated versions appeared in U.S. newspapers with local place names substituted for the original British settings—doubly infuriating the author, who took great relish in detailing the destruction of London and its suburbs. Following his ascent to fame, Wells befriended Lowell and promoted Lowell's theories about Mars in many articles.

After *War of the Worlds*, Martian fever heightened in 1911 when Chicago businessman Edgar Rice Burroughs penned the adventure tale "Under the Moons of Mars." Burroughs, who followed this novel with *Tarzan of the Apes*, later pretended to be embarrassed by his success as a writer of adventure literature. He was a frustrated man of action, with an upper middle class background. As a youth he had been thrown out of several private schools, but shone at the Michigan Military Academy, where he was known for his horsemanship, leadership, and pranks. Failing his exams to enter West Point, he had a short stint teaching at the Michigan Military Academy then joined his brothers' efforts at ranching and a gold dredging operation in Idaho. After that, for several years, he signed on with the U.S. Cavalry to patrol the Southwest where, he later remarked, he and his fellow soldiers were fortunate never to run into any Apaches.

After marrying Emma Centannia Hulbert, a young woman from his wealthy Chicago neighborhood, Burroughs established himself as an able manager of the stenography department at Sears, Roebuck and

Company. Bridling at the conformist pressures of middle management white-collar work (a talented cartoonist, he dashed off numerous satirical sketches of his employers and employees), he left Sears to launch a series of failed businesses. The last of these included pencil sharpener sales, and the dispensing of business advice via mail to fellow frustrated entrepreneurs to whom he would offer the precepts of efficiency. Finally, while working for his brother's stationery company, during idle hours he began scrawling the ideas for a Mars novel on the back of discarded letterhead. The ensuing saga, signed "A Normal Bean" but attributed to "Norman Bean," was serialized as "Under the Moons of Mars" in the pulp magazine *All-Story*, and later became novelized as *The Princess of Mars*.

*The Princess of Mars* relied (albeit loosely) on the template for Mars that Schiaparelli, Flammarion, and Lowell had developed. Burroughs specifically credited Flammarion's novel *Uranie*, as an inspiration. Burroughs's Mars, or Barsoom, as its inhabitants called it, was a dying planet, crisscrossed with canals, and populated by many humanoid species of different hues—all with a penchant for battle. More importantly, in the novel Burroughs introduced his hero, perhaps more accurately, his superhero, John Carter.

Carter, a former Confederate soldier, is prospecting in the Southwest when Apache Indians chase him into a sacred cave. No spaceship is necessary to transport him to Mars. Facing imminent death, he leaves his body behind and journeys to Mars, using the method of spirit-travel common to Flammarion's speculative dream journeys through the cosmos. On Mars, Carter discovers he has superhuman fighting powers, thanks to the planet's weak gravitational pull. (The creators of the comic book *Superman* also would rely on this scenario for explaining their hero's great strength when arriving from planet Krypton to the wimpy gravitation of the Earth.)

On Barsoom, Carter is captured by the Tharks, barbaric green-skinned six-limbed nomads nearly twice his height—modeled loosely on the warrior-like Apaches of the Southwest that he had fled. He soon becomes their champion via gladiatorial competition, and then rescues the glorious Heliumite, Dejah Thoris, a red-skinned Martian princess before her forced marriage to a crude member of a conquering Heliumite faction. Carter defeats her people's enemies, weds Dejah (who lays eggs that eventually hatch into offspring), and ultimately rules the planet. Again emphasizing the image of Mars as a dying planet, the book ends with Carter trying to gain access to and so repair a gigantic factory that replenishes the planet's dwindling atmosphere. (How? Burroughs craftily asserted that the atmosphere generator relied on the "ninth ray" of the color spectrum unique to Barsoom.)

With his continuing tales of John Carter on Mars, Tarzan of the Apes, and the weird world of Pellucidar beneath the surface of the Earth, Burroughs helped make pulp magazines an important element of popular culture for the next half-century. While Wells was certain of his own genius, and picked fights with other literary luminaries, including Henry James, Burroughs was content to be regarded as a writer of adventure fiction. He told one interviewer, "I don't think my work is 'literature,' I'm not fooling myself about that." He once even suggested his efforts were "in the same class with the aerial artist, the tap dancer, and the clown."[16]

Publisher Hugo Gernsback was another critical literary convert to the cause of spaceflight fantasy. He claimed that at age ten, while still living in Luxembourg, he had come across a copy of Lowell's book *Mars*. It inflamed his imagination. He fell into a fever and for several days hallucinated about the fantastic beings and technology that would be discovered on Mars. In 1926, Gernsback, then a hawker of radio parts in the United States, and owner of Experimenter Publishing, launched *Amazing Stories*—a pulp magazine that promoted stories of "super science"—and its pages were soon populated with Buck Rogers and other space cadets, BEMs (frequently from Mars), "people-like" aliens, and truly alien aliens. Gernsback helped create a market for science fiction and a new generation of science fiction fans and writers.

* * *

Although Lowell's camp was suffering defeat by 1910 in scientific circles, the belief that there was vegetation (perhaps a form of lichen) on Mars remained common through the 1940s. Likewise, the Lowellian notion of Martian canals—those prodigious feats of engineering—once enshrined in maps, globes, and schoolbooks only slowly faded. By 1913, most astronomers refuted their existence, but the canals remained a staple of pulp and higher-minded narratives (as did theories about Mars's oceans and vegetation) into the 1960s. As to Mars's dissipated waters—Flammarion and Lowell had the right idea. The Curiosity rover has shown that the planet once did hold large rivers as well as lakes. More recent science fiction and futurist tracts no longer concern themselves with encounters with Martian societies. Instead they involve the human project of settling and "terraforming" the planet (an idea that began in the 1940s) and biologically altering human settlers (or "aeroforming") to adapt to life there—in the process becoming the Martians always depicted as potential doubles.

It took a century beyond the publication of H. G. Wells's *The War of the Worlds* for NASA to land its first unmanned rover on Mars, but as Wells commented in the early twentieth century, he and his peers had effectively conquered Mars through their imaginations, turning scientific data into the stuff of myths. Pure science was not necessary to create enduring interest in space exploration. Fifty years after Wells's pronouncement, Ray Bradbury noted, "It is part of the nature of man to start with romance and build to a reality."[17] But science fiction writers only slowly conformed to new facts. Lowell's vision was more fun than that of Green and other debunkers. The concept of a "dying planet" and a super-race carving out grand canals was too rich to let go of entirely, even in the post-Sputnik era. In a preface to a book for young readers, *The Secret of the Martian Moons* (1963), Donald Wollheim noted that he chose "to depict Mars as presented by the late Professor Lowell, of Flagstaff Observatory . . . It has always been the view most exciting to men's minds." Wollheim added, in a bit of a stretch, "and it is still upheld by a substantial section of planetologists."[18]

The Mariner 4 flyby in 1965 suggested a planet barren of vegetation or any traces of Barsoomian civilization. As late as 1967, astronomer Carl Sagan, who had Barsoom vanity license plates on his car, was still a hold out. In a *National Geographic* article he proposed a turtle-like creature as one of the many possible inhabitants of the Red Planet. In 1972, Mariner 9 offered high resolution images mapping much of the planet's blasted surface—a surface conspicuously lacking lichen or Flammarion's red vegetation. Sagan gave up on the idea that Mars might be Barsoom. Honoring data above myth, he expressed delight at evidence of ancient riverbeds, and posited that the former reports of seasonal blooms most likely could be explained by dust storms that periodically uncovered darker rock on the surface.

And were Robert Zubrin and his opponent John Rummel of SETI oddballs to passionately debate issues of planetary contamination before an audience in 2018? Hardly. In September 2019, NASA's chief planetary scientist Jim Green suggested that it was more than likely that with upcoming rover missions and soil sample returns to the Earth, evidence of past or even current Martian life would soon be found. "I think we're close to finding it and making some announcements."[19] Meanwhile, as of this writing, robots continue their methodical survey of Mars, traversing a landscape that honors both sides of the canal debate, with craters named after Flammarion, Schiaparelli, Lowell, Antoniadi, and Green, as well as H. G. Wells, Kurd Lasswitz, Edgar Rice Burroughs, and later science fiction writers.

# CHAPTER TWO

# ROCKETEERS

• • •

"All night long the Martians were hammering and stirring, sleepless, indefatigable, at work upon the machines they were making ready . . ."
—H. G. WELLS, *War of the Worlds*, 1898

"God pity a one-dream man."
—ROBERT GODDARD, 1909

In the twenty-first century, space launches have become commonplace. NASA, Roscosmos (the Russian Space Agency), the China National Space Agency, the European Space Agency (ESA), the Japan Aerospace Exploration Agency, and private companies such as SpaceX, Blue Origin, United Launch Alliance, and others, all loft spacecraft, while France, South Korea, India, Israel, Italy, Australia, New Zealand, and Iran also have launch-capable space programs. There were over one hundred launches in 2019, and this number will increase once rockets are used for space tourism and, possibly, for global commercial travel—in 2017 Elon Musk remarked that SpaceX could arrange thirty-nine-minute rocket flights from New York to Shanghai. While spacefaring enthusiasts claim one of their goals is to preserve the Earth's environment, experts suspect that fuel exhaust from such launch activity—even if it does not add greatly to global warming—might damage the ozone layer that shields the Earth from ultraviolet radiation.

Having to negotiate such a problem would probably have delighted early rocketeers, who could only dream of a future where spaceflight was commonplace while they watched their test rockets blow up or

veer sideways and terrorize local livestock. Following Ray Bradbury's
dictate that great projects "start with romance and build to a reality,"
the blueprint for the first space launch was not a product of NASA
or the Soviet Space Agency, but sprang from the noble brow of Impey
Barbicane, president of the Gun Club of Baltimore, as presented in
Jules Verne's *From the Earth to the Moon* (1865). The novel begins
after the Civil War has ended and members of the Gun Club of Baltimore, arms manufacturers, are "reduced to silence by this disastrous
peace."

Bored, desperate, losing interest in purely speculative designs for
gunnery, the Gun Club nearly dissolves until its president Impey Barbicane proposes a "grand experiment worthy of the nineteenth century."
He suggests that as there were, unfortunately, no wars imminent, they
use their knowledge of gunnery to reach the Moon (or, as he calls it,
The Queen of the Night). "The means of arriving thither are simple,
easy, certain, infallible. . . . I find that a projectile endowed with an
initial velocity of 12,000 yards per second, and aimed at the moon,
must necessarily reach it."

When the president's speech concludes, the crowd roars with delight
and carries him in a torchlight parade through Baltimore's streets. The
Gun Club sets to the task, and, eventually, Barbicane, two other brave
voyagers, and two dogs enter an aluminum capsule that is lowered
into the Columbiad, an enormous cannon, in Florida, and, following a
recipe common to many circus acts, they are blasted into outer space.
Decked out in Victorian garb, dining on soup, beefsteak, and bottles
of Burgundy (microgravity, apparently, had not yet been invented),
Verne's voyagers manage (in the novel's sequel *Round the Moon*) to
orbit the Moon and eventually return to Earth.

Tales by Verne, H. G. Wells, and others quickly inspired real life
Barbicanes in Russia, Germany, and the United States. The "origin
story" of the space rocket, however, is muddied: Verne was certainly
not the first to imagine space travel, and, depending on nationalistic
allegiances, Robert H. Goddard of the United States, Konstantin Tsiolkovsky of Russia, or Hermann Oberth of Austria-Hungary can be

credited for developing modern rocketry. Soviet propagandists ret-rospectively made of Konstantin Tsiolkovsky a "people's hero" and recast his life in saintly terms, piling on achievements to indicate he had thought through all the modern rocket's design specifications; nevertheless, it is safe to say of the "big three" Tsiolkovsky, who never actually built or tested rockets, overcame the greatest odds to breathe life into spaceflight.

Tsiolkovsky was born in 1857, the fifth of eighteen children, in a small village, Izhevskoye, southeast of Moscow. His father was a for-ester and would-be inventor; his mother, whom his father claimed pro-vided the family's "spark of talent," was from an artisanal family. At age nine, after a winter accident (depending on the account, the young inventor-to-be was tobogganing or ice skating), Tsiolkovsky con-tracted scarlet fever, and as a result became deaf for life. This disability ended childhood as he'd known it. "I often became and behaved awk-wardly among the children my age, and among people in general . . . I felt isolated, even humiliated . . . an outcast from society," he wrote, and "this caused me to withdraw deep within myself, to pursue great goals."[1] Books and dreams helped him escape. He was a fan of Jules Verne, yet he insisted his interest in outer space predated his reading of scientific romances. He later wrote that from "very early childhood, before I ever read books, was a dim consciousness of a realm devoid of gravity, where one could move unhampered anywhere, freer than a bird in flight. . . . I dimly perceived and longed after such a place unfettered by gravitation."[2]

At age thirteen, after his mother died, Tsiolkovsky dropped out of school, but continued to study math and science, relying on texts from his older siblings and from his father's small science library. At age sixteen, his father encouraged him to leave the provinces for Moscow, where, with the help of a metal ear trumpet he sat in on university lectures. His father sent him a small stipend of ten to fifteen rubles a month, which he used primarily to buy sulfuric acid, glass beakers, and other materials to conduct experiments in chemistry and physics. "I remember very well that I had nothing to eat but brown bread and water. For all that, I was happy with my ideas, and my diet of brown bread did not damp my spirits."[3] He also discovered an unofficial university, the Rumiantsev Library in Moscow. Inside this edifice bedecked with Greek columns, the city's first public library, Tsiolkovsky met the librarian and philosopher Nikolai Fedorov. Fedorov admitted Tsiolkovsky into the circle of young scholars he mentored. Each day at his desk, Tsiolkovsky would find piles of useful books. Fedorov also provided him with clothing and food.

To understand Tsiolkovsky's aims and achievements, it is worthwhile to survey Fedorov's thought, which heavily influenced his pupil. Fedorov, who led an ascetic life, dedicated to Russian Orthodox practice, was a lead voice in a philosophical movement in Russia later dubbed "cosmism"—which has had widespread influence in Russian literature, arts, and science. Cosmism, at its simplest, proposed that the Earth must be understood as part of a larger cosmic system. Its central tenet was that evolution is an active force; humanity could reshape itself as life spread, inevitably, into interplanetary space. While cosmism embraced a vision of scientific advance and progress, its participants rejected the mechanistic universe proposed by European Enlightenment science. Instead, cosmism presumed a lively universe, endowed with mystical purpose and connection.

Fedorov was the illegitimate son of Prince Pavel Ivanovich Gagarin, a bon vivant who founded theater companies, sired many offspring, ultimately ran through his fortune, went bankrupt, and was rejected by his wealthy kin. Federov, nevertheless, received a superior education. He quit school after a furious argument with one of his teachers during oral exams. For many years he was a schoolteacher, dedicated to his students, but fired from most of his posts because of his insistence on simple, unwavering principles, which led him to condemn other teacher's harsh treatment of students. When he had money, he gave much of it away to needy students. He dressed shabbily, preferred to eat simple vegetarian meals when he bothered, gathered no possessions, wore a thin and ragged coat year-round, lived in plain rooming houses, and slept only a few hours a night on top of a "humpback" trunk.

His force of thought and ascetic lifestyle brought him disciples, including, for a time, Leo Tolstoy, who regarded Fedorov as a father figure. Fyodor Dostoyevsky also was intrigued by Fedorov and threaded his ideas throughout *The Brothers Karamazov*. Eventually, one of Fedorov's disciples secured him the position at the Rumiantsev—where he thrived. As a librarian Fedorov functioned much as he had as a teacher, adding useful books to patron's requests, steering research, helping the indigent, and inviting scholars to his office for discussions.[4]

A religious thinker, steeped in Russian Orthodox belief, Fedorov was no scientist, but a metaphysician. He believed that the "Common Task" of humanity, which united all, was to defeat death. In the soft version of this philosophy, as expressed in Fedorov's 1892 short story "Karazin: Meteorologist or Meteorurge?" humanity needed "to marshal all its current resources and seek additional ones in order to resolve its differences with the blind, intractable forces of nature . . . nature, precisely because it is intractable, is an agent of death."[5] In this context, he sounds like an heir to the Enlightenment, urging that reason overcome ignorance and the resistance of the material world. Fedorov saw death as emerging in two manners, from "disintegration"— social, psychological, and physical, or in the tendency to "fusion," the conglomerating of humans in larger social institutions that deadened their souls. The main purpose of living was to overcome death, to resist both disintegration and fusion, and to live as if an immortal, restoring wholeness.[6]

But Fedorov's "hard" program insisted that humanity must literally defeat death. His stated goal, reflecting a primal version of Christianity, was to resurrect the dead and pursue immortality. He believed the Western vision of progress was useless, merely accelerating "fusion," as kinship and feudal ties gave way to the needs of the lifeless collective of "society." In his ideal universe there would be no sex, no birth, and no death. Scientists would find ways to extend human life. The formerly dead would be, via breakthroughs in science and medicine, resurrected and so join the "living sons." We would no longer (metaphorically) cannibalize our dead, we would revive them. Organizations such as the military would disband as mankind realigned to pursue its "common task."

Earth, alas, would get crowded. To prepare for the coming resurrections, Fedorov wrote that, "The conquest of the Path to Space is an absolute imperative, imposed on us as a duty in preparation for the Resurrection. We must take possession of new regions of Space because there is not enough space on Earth to allow the co-existence of all the resurrected generations."[7] Travel into the cosmos might also be necessary to recover traces of the long dead, to be reassembled in new, posthuman forms to settle the universe. There was no point waiting for the apocalypse. We were in charge—the dead must rise again and heaven emanate outwards. Fedorov speculated that humans even might learn to manipulate the Earth's electromagnetic fields, freeing the planet from the Sun's gravity, so that the Earth, now a liberated spaceship, could set course for new areas of the universe.[8]

Tsiolkovsky later noted of Fedorov, "It is no exaggeration to say that for me he took the place of university professors."[9] Yet Tsiolkovsky insisted that he and the librarian never discussed space travel. This is open to debate. One of Tsiolkovsky's contemporaries, Viktor Shklovsky, described a scene in which Fedorov approached the young student, realizing he was deaf, and wrote on a paper: "I'm going to do mathematics with you, and you'll help mankind build rockets so that we will finally be able to know more than the earth."[10] But in his memoirs, Tsiolkovsky insisted that spaceflight was his own private mania.

Indeed, shortly after arriving in Moscow, Tsiolkovsky was speculating on ways to launch humans into space, and hit first upon the notion of centrifugal force. He imagined a device like a giant flywheel that could shunt one off into the stars. (The juvenile Robert Goddard also later came upon and rejected this concept.) Spaceflight grabbed the young Tsiolkovsky's imagination, so that after envisioning the great flywheel, he said he was unable to sleep and "wandered about the streets of Moscow, pondering the profound implications of my discovery."[11] He spent three years in Moscow, under the sway of Fedorov and his circle of young scholars, reading books, attending lectures, and conducting experiments in his lodgings. Hearing stories of his poverty, his father urged him to return home.

After his apprenticeship in Moscow, Tsiolkovsky passed a teaching exam and began a long career as a teacher of mathematics and science in Russia's provinces—his first position was in a town where Fedorov had once taught. In 1880, he married a preacher's daughter, Varvara Sokolova, and in 1893 he took his second teaching post in Kaluga where he remained the rest of his life. Although some locals regarded him as a crank, or as the "Kaluga eccentric" (*kaluzhkii chudak*, his nickname), Tsiolkovsky was a dedicated and popular teacher. In this respect, he emulated Fedorov's earlier career in teaching. Often he would allow students into his home workshop, or invite them to the second floor balcony where they would open the "door to outer space" and stargaze with Tsiolkovsky's telescopes.[12] In his spare time he wrote educational pamphlets and gave lectures.

Tsiolkovsky also continued to study and experiment on his own and corresponded with members of the Society for Physics and Chemists. Dimitri Mendeleyev, the developer of the Periodic Table of Elements, let him down gently when, in 1881, Tsiolkovsky sent the society his paper "Theory of Gases," which proved laws concerning the kinetic behavior of gases that were already well-established. Tsiolkovsky next wrote a paper called "Free Space" in 1883, in which he deduced how ships might move in outer space, free of gravity. With a grant of 480 rubles from the Russian Academy of Sciences, in 1899 he built a wind tunnel, and began to test designs for an ideal dirigible. He also, apparently, designed a monoplane in the 1890s that later proved flight-worthy. But his true dream was to enable spaceflight.

In 1898 Tsiolkovsky wrote "The Exploration of Cosmic Space by Means of Reactive Devices." In the paper, published in 1903, he noted that the concept of a rocket gun, à la Verne, was unworkable. Such a gun would have to be 300 meters in length, that is, as tall as the Eiffel Tower, and withstand terrific pressures to remain intact. Likewise the initial explosion of the device, providing one thousand times ordinary gravity, would make it impossible for a living organism or any sensitive instruments to survive intact. Even if they survived, it would likely be a one-way trip.

Instead Tsiolkovsky proposed a rocket, or reactive device, relying on liquid fuels such as hydrogen, oxygen, kerosene, or alcohol that would gradually increase the rocket's speed to prevent overheating as the vehicle passed through the Earth's atmosphere. The engine would include a "jacket" in which liquid metal could circulate as a coolant—as might the liquid gases used as propellant. The rocket would have automatic controls to set the rocket's course and prevent it from rotating. Most significantly, the rocket could be a vehicle for humans. The front of the rocket would be "equipped with electric light, oxygen, and means of absorbing carbon dioxide, odors, and other animal secretions; a chamber, in short, designed to protect not only various physical instruments but also a human pilot."[13] He added, "In many cases I am limited to guesses or hypotheses. I am not deluding myself and I am perfectly aware that I am not solving the problem in its entirety, that a thousand times more work than I have done must be invested in its solution."[14]

He calculated the escape velocity necessary to leave the earth's atmosphere, and with the Tsiolkovsky Equation established the maximum velocity of a rocket based on the exhaust velocity of the fuel and the relative mass of the rocket before and after fuel use. He argued that a reactive rocket could be worthy of interplanetary travel. He also envisioned an early space station: "given an oblique ascent it is possible to construct a permanent observatory that would travel for an indeterminate length of time around the Earth, like the Moon, beyond the limits of the atmosphere."[15]

Interest in Tsiolkovsky's 1903 paper was virtually nil. It was not translated. In 1912 he updated the paper and speculated that radioactive materials might be used in the engines. He also sketched out the benefit of establishing Earth-orbiting space colonies as a first step toward expanding civilization deep into the universe. Imagining a technology well in advance of that of his own era, he remained a

marginal figure until after the Bolshevik Revolution, when Soviet officials refashioned the brilliant crank into a national treasure.

Tsiolkovsky complemented his scientific studies by writing science fiction stories. These stories, generally, dispensed with character
development and instead described as realistically as possible life off
planet Earth. His tone was gentle and didactic, making this work ideal
reading for the young students that he hoped to inspire. In his first
published story, "On the Moon" (1893), he described a Flammario-
nesque dream voyage to the Moon. His characters—a narrator and
a "physicist"—find themselves in a house transported to the Moon.
The physics instructor helps explain their unique situation, while the
narrator introduces his readers to what it would be like, for example,
to bound across the Moon's low gravity surface, while successfully
chasing the Sun during the sunset. It also includes macabre details, as
for example, the Moon visitors' horse panicking, racing off, and then
shattering against a rocky mass: "the meat and blood froze at first and
then dried out."[16] This story would not be the only time Tsiolkovsky
used the motif of a shambling Earth house in outer space—not for
verisimilitude, but to imagine what would happen to such materials,
and, likely, as a symbol of his deep belief that humanity's true "home"
was in outer space.

His longer novel *Dreams of Earth and Sky and the Effects of Universal Gravitation* (1895) included as a stand-in character, the "Gravity
Hater." The book lets the Gravity Hater vent about his distaste for this
cosmic force, but also describes what would happen to life on Earth
if gravity disappeared. Most importantly, it depicts how to build an
orbiting space station—among his proposals, to convert solar to electric power. Of his science fiction, Tsiolkovsky noted, "Many times I
essayed the scientific concept through the task of writing space novels,
but would then wind up becoming involved in exact compilations and
switching to serious work."[17] This science fiction, ultimately, would be
as important as his efforts at theory, as he wrote many pamphlets and
stories that inspired the next generation of rocket researchers in the
Soviet Union.

In addition to his studies of rocketry and science fiction, Tsiolkovsky indulged in philosophy. In later essays such as, "Citizen of the Universe," he detailed a monist theory, offering his belief that all matter contained the seeds of life, i.e., the minute, indestructible, fully-alive "true atoms." (All the easier, perhaps, to gather and resurrect earlier combinations.) The need to approach perfection led him to unsavory speculations—to him, not all combinations of "true atoms" appeared equal.

"I do not desire," he wrote, "to live the life of the lowest races [such as] the life of a negro or an Indian. Therefore, the benefit of any atom, even the atom of a Papuan, requires the extinction also of the lowest races of humanity."[18] While Fedorov insisted all must be included in his resurrection, Tsiolkovsky saw evolution as a process of weeding out the unfit, leaving those who could approach perfection and eventually justly rule. In other essays Tsiolkovsky discussed the abundance of life in the universe and proposed that we had not been contacted by higher intelligences from other planets, star systems, or galaxies because we were not ready. Earth could be thought of as a backwater preserve where humans, left alone, might yet come up with good ideas.

Tsiolkovsky's importance in his home country increased after the Bolshevik revolution. In 1918, he wrote a letter to the newly-formed Socialist Academy of Social Sciences, describing his situation as an innovative scientist in the provinces, forced to teach school for decades, his work ignored by the tsarist elites. The letter resonated; a month later he was granted a monthly stipend, admitted to the Soviet Academy, and encouraged to popularize his ideas through pamphlets, lectures, and interviews. By the mid-1920s, a space craze was blossoming in Russia, with the public awash with hopes of the future and the promises of science, and Tsiolkovsky, sprung from the people, a self-taught genius, became an idealized emblem of socialist progress.

• • •

In 1899, the year after Tsiolkovsky published "The Exploration of Cosmic Space by Means of Reactive Devices," approximately 4,400 miles away, in Worcester, Massachusetts, the seventeen-year old Robert Goddard climbed a cherry tree on his aunt's farm, armed with a saw and hatchet to trim dead limbs. He had recently read, serialized in the *Boston Post*, a pirated version of H. G. Wells's *War of the Worlds*, as well as Garrett P. Serviss's *Edison's Conquest of Mars*, both of which inflamed his imagination.

Busy in the cherry tree, Goddard was thinking first about his plans for a "frog hatchery" (he was an adolescent of many interests), when, looking out at the fields from the tree limbs he was struck with the desire to create "a device which had even the *possibility* of ascending to Mars." His first impulse, similar to Tsiolkovsky at his age, was to imagine a centrifugal device that could whirl a craft off the Earth. "In any event, I was a different boy when I descended the tree from when I ascended, for existence at last seemed very purposive."[19] Although this story has an apocryphal feel, it is undoubtedly true, as Goddard proceeded, for many years, on October 19, to celebrate its anniversary with returns to the cherry tree, all noted in his journals. (In 1901 he wrote "walked down to village and climbed cherry tree in morning." In 1909, "Cherry tree anniversary in evening." In 1917, "Leaned hard against the trunk of the tree." In 1922, "Went over to old house.")

Goddard's passion for space travel slowly overtook other interests. Bouts with illness earlier had turned his career interests to medicine. But he also had a passion for invention, a passion no doubt fed by "Edisonades"—the genre of turn-of-the-century books for children in which Edison or a boy inventor performed heroics—even, in the case of *Edison's Conquest of Mars*, taking revenge on Wells's Martians, by arming a spaceship crew with an antigravity device and a disintegrator ray.[20] As for Goddard's early inventions, as a child, in addition to the frog hatchery, he attempted to build an aluminum balloon filled with helium, and to manufacture artificial diamonds—this experiment led to a violent explosion that drove glass tubing up through the ceiling into the attic. His only injury was a cut knuckle from a flying shard. Rocketry, intentional or not, was slowly to dominate his interests.

Soon after leaving the cherry tree, Goddard spoke with an older student of his plans for a centrifugal space launcher and was told it would not work. He didn't accept this at first and began to build various models, and it slowly dawned on him that "If a way to navigate space were to be discovered—or invented—it would be the result of a knowledge of physics and mathematics."[21] Overcoming his own initial distaste for mathematics, he applied to Worcester Polytechnic Institute. While there he began filling green notebooks with speculations about spaceflight. He explored "using the magnetic fields of the earth, shooting material to a 'spaceship' by means of electric or other guns, an airplane operated at high speed by the repulsion of charged particles," as well as "artificially stimulated radioactivity. . . . propulsion in space by repulsion of charged particles. . . . the use of liquid propellants, and several other plans."[22]

Rejecting the Verne gun concept, he thought instead of "raising an explosively propelled apparatus to a great initial height by balloons"; he also wondered if there was a way of gaining traction by "reaction against the ether" then believed to fill space, and made notes such as the value of "circling a planet to decrease speed before landing."[23] He kept filling up the green notebooks, even after he graduated from the Polytechnic and moved on to Clark University where he gained a PhD in 1911, became a fellow, then later, professor.

Despite serious bouts of illness with stomach complaints, kidney problems, and lung disease, he continued refining his approach to his big dream of spaceflight. (Ignoring, in the process, the advice of his journal entry from 1909: "God pity a one-dream man.")[24] Just as Tsiolkovsky had worked in isolation with little to no institutional support for most of his life, Goddard is often depicted as someone working outside the norms. Although he preferred to work with only a small support crew, unlike Tsiolkovsky, Goddard was very much a member of the scientific elite and he and his assistants benefited from numerous grants, including funding from the Smithsonian, the U.S. military, and the Guggenheim Foundation.

Goddard left behind his early experiments in electronics (hoping to discover a way to create an engine that would pulse against the "ether," now described as ion propulsion) and shifted to research on chemical propellants when at Clark, and then as a research instructor at Princeton. He inadvertently breathed in sulfuric acid fumes during an experiment and this forced him to return home to convalesce. He soon recovered. In 1914 he filed a patent for a method to liquefy gases to be used as rocket fuels, and the following year two patents for rocket engine designs. When his funding from Clark University proved not enough, he applied to the Smithsonian Institution in 1916, proposing that his rocket could be used for meteorological research in the atmosphere's upper reaches. He was awarded $5,000. (This was significant—adjusting for inflation, it would translate to about $115,000 in 2018.)

Within days of the United States entering World War One, Goddard wrote to the secretary of the Smithsonian Institution arguing that military applications for his rockets should be explored. The secretary responded that the military would be much more interested after a successful test flight. Goddard's main technical difficulty was the rapid reloading and firing of cartridges of chemical explosives to power a rocket that could leave the atmosphere. After the Smithsonian's objection, Goddard directly wrote the army Chief of Ordnance insisting his rocket had "possible applications to warfare."[25]

Slowly word of his research circulated. In late April 1918, an army ordnance specialist recommended that Goddard be allotted $10,000 toward secret work on the new weapon. Two months later, Goddard moved to California where he could work in secret outside Pasadena near the Mount Wilson Solar Observatory. On the train from Chicago to Los Angeles, his suitcases were packed with explosives.

In Pasadena, he worked with several assistants and machinists on his "rocket gun" (one outcome, many years later, was the bazooka). The work did not always go smoothly. He wrote, "This morning, while

Mr. Hickman was removing the paper cap from the blasting cap on a 10-gram cartridge, the cartridge exploded, and necessitated the amputation, on the left hand, of the thumb to the first joint, and the first two fingers to the second joint; and on the right hand, of the first finger to the first joint . . . Fortunately his eyes escaped any injury."[26]

Clarence Hickman, a graduate student of Goddard's at Clark, later said much of his hand was saved because "the resident surgeon at the hospital was not available, but they got a man who works for the railroad company. To them every joint of the fingers is worth so much money, you know."[27] Despite the loss of many finger joints, he continued to work with Goddard and went on to a successful career as a designer and physicist, later working at Bell Laboratories, and advising the military on weaponry during World War Two. (An amateur clarinetist, he created one he could play despite his disabilities, and he also made improvements to the player piano for the American Piano Company).

Life in California also had its light moments. Goddard wrote of a Sunday hike he took into the hills "Very pretty . . . Came down at sunset . . . Lights in Pasadena came out. Stopped at Observatory and saw Mars. 'Any signs of life?' 'No.' A soldier said, 'Any signs of life in Pasadena? No, never was and never will be.'"[28]

By the war's end Goddard, Hickman, and assistants designed several functioning rocket guns, one to be affixed to an airplane as heavy artillery, and another a recoilless anti-tank weapon, which would form the basis of the bazooka. A working version was demonstrated a few days before Armistice. In the ensuing peace, the army immediately ended Goddard's weapons research. Following the war, a Massachusetts newspaper printed an article about Goddard's wartime work, "Invents Rocket with Altitude Range 70 Miles," with subhead "A Terrible Engine of War Developed in Worcester by Dr. Robert H. Goddard." As rumors circulated, Goddard convinced the Smithsonian to publish, in December, 1919, "A Method of Reaching Extreme Altitudes," as a booklet. It described a multistage rocket to be used for meteorological readings and a flight to the Moon. The rocket would be unmanned. He suggested that "flash powder" could be exploded on the Moon, so as to be visible from the Earth and prove the rocket had left the Earth's atmosphere.

In January, 1920, the *Boston Herald* printed "New Rocket Devised by Prof. Goddard May Hit the Face of the Moon," and numerous papers took up the story, including the *New York Times*. Much merriment followed. The Bronx Exposition company offered the Smithsonian use of their amusement park as a launch pad for a Moon rocket. Another writer expressed concerns to the Smithsonian that Moon inhabitants might interpret a rocket launch as an act of war and retaliate. A few weeks later, on page one of the *New York Times*, Claude Collins, apparently a captain in the New York City Air Police, misunderstanding news about Goddard's rocket, announced, "Believing the plans of a noted scientist to send a super-rocket from the Earth to Mars, in the body of which a person would be stationed, can be developed into a reality, I hereby volunteer."[29]

Trying to evade what he called in a later report to the Smithsonian "sensationalism," Goddard published a general statement stressing that there had been too much emphasis on his "flash powder experiment," let alone other unspecified "interesting possibilities" connected to rocketry, and not enough on atmospheric research. He proposed a public subscription to support his work. Despite his worries over sensationalism, in a March 1920 report to the Smithsonian, Goddard outlined how liquid oxygen and hydrogen fuels would make possible journeys to planets "without an operator" and "with an operator." Efficiency could be improved if, on the Moon or planets, using solar energy, more hydrogen and oxygen could be separated and liquefied for the return voyage to Earth.[30]

While Goddard gained few colleagues in the United States, his work drew attention abroad where other physicists and aeronauts shared his fascination with spaceflight. Two of the most prominent: French aircraft designer Robert Esnault-Pelterie, and Hermann Oberth, a Romanian physicist, requested copies of his articles and patents. Oberth's ensuing book *Die Rakete Zu Den Planetenraumen* (1923), or, *The Rocket into Planetary Space*, not only discussed rocket designs but took head-on what Goddard shied away from—at least in public. Oberth, a nimble mathematician, had no hesitation in stating that humans could traverse interplanetary space. He noted, for example in the book's introduction, that "These machines can be so constructed that men can be lifted in them, apparently with complete safety."

Oberth also projected the future of space exploration, and possibilities such as establishing space stations, and placing giant mirrors in space to serve as weapons or agents of climate change. His ultimate goal, stated in a later work, was "to make available for life every place where life is possible. To make inhabitable all worlds as yet uninhabited, and all life purposeful."[31] In an addendum to *Die Rakete*, Oberth offered a summary of Goddard's work, but in such a way as to imply his own preeminence in the field, singling out, for example, the American's seeming preference for solid propellant, while Oberth championed liquid propellants such as alcohol, liquid oxygen, and hydrogen.

Goddard, who had long experimented with liquid propellant—and would be the first to launch such a rocket—did not engage Oberth in public debate. But to his sponsors at the Smithsonian, he commented, "I am not surprised that Germany has awakened to the importance and the development possibilities. . . . and I would not be surprised if it were only a matter of time before the research would become something in the nature of a race."[32]

Meanwhile, Goddard continued his work on liquid propellant rockets. While the U.S. press never took Goddard too seriously, describing recent failures of his "moon rocket," in Europe, the reaction was quite different. In 1928, the German press reported that Goddard had plans to launch himself in a rocket to the Moon.[33] Oberth's book, news and rumors of Goddard's launches, the lectures of enthusiasts such as Robert Esnault-Pelterie, and early science fiction films opened up a rocket craze in Europe in the 1920s that reflected the greater frenzy over aviation—in which pilots such as Charles Lindbergh and World War One aces on both sides became popular heroes.

• • •

Nowhere was the space fad stronger than in the Soviet Union where a taste for modernity, an official policy of atheism, and an embrace of technological utopianism mixed uneasily with cosmist aspirations. In the Soviet Union, science lecturers abounded, and popular science articles were widely distributed. Along with fascination for the foreigners Goddard and Oberth, Tsiolkovsky was "discovered" and became a people's hero. Alexei Tolstoi's science fiction work *Aelita, Queen of Mars* was published in 1923, and made into a movie in 1924, the year that Mars and Earth were in the closest opposition in a century. In the novel *Aelita*, a Soviet inventor, Los, and Gusev, a Red Army officer, travel to a dying Mars on an egg-shaped spaceship, and learn that many of its peoples are descendants of dwellers in Atlantis. During a workers' revolution on the Red Planet, which Gusev helps lead, Los and the semi-human Martian Queen Aelita fall in love and attempt double suicide in a cave. Only Aelita dies. A mournful Los and Gusev escape back to Earth.

The Soviet taste for Mars and beyond dovetailed with new movements in Soviet arts and letters. Artist Kazimir Malevich wrote to a friend that "an aspiration towards space is in fact lodged in man and his consciousness," and dubbed himself, in 1917, "General Secretary of Space." In 1920, Malevich launched the art movement Suprematism, based in abstract geometric forms. Malevich sought to free art from the "dead weight of the real world" just as Tsiolkovsky longed to be free of gravity. Suprematists would move to a "new realism in painting, to objectless creation." This abstract art movement paralleled interest in space. Malevich installed a telescope in his art studio, and proposed a way to reach the Moon: "Travel along a straight line toward any planet can only be accomplished by means of intermediate Suprematist satellites in circular motion, which would form a straight line of rings from satellite to satellite." Suprematist renderings of satellites and other abstractions were meant to allow the viewer to directly experience a space freed from the Earth and other mediated categories of the real.[34]

In 1924, an amateur society of Soviet rocketry enthusiasts formed, and its members, largely students and workers, promoted numerous lectures. Rumors of rocket launches to the moon led to a near riot at one such meeting in Moscow in October 1924. A *Pravda* article from that year indicated, "Within a few years, hundreds of heavenly ships will push into the starry cosmos."[35] From 1923 to 1932, 250 articles and thirty books appeared in the Soviet Union on spaceflight, whereas in the United States, in that same period, only two books on the subject appeared.[36]

Fascination with spaceflight continued beyond the Lenin era. Three years after the Bolshevik leader's death, a group of Tsiolkovsky aficionados, diehard cosmists, mounted "The World's First Exhibition of Models of Interplanetary Apparatus, Mechanisms, Instruments, and Historical Materials," in Moscow in 1927. The exhibition included a lunar panorama, models of rockets, and a shrine to (the still living) Tsiolkovsky, that included photographs, models of his airships, and samples of his writings. In two months, the exhibition gained at least 10,000 visits from schoolchildren, workers, the Futurist poet Vladimir Mayakovsky (already soured on the revolution) and other avant-garde artists. The exhibition's comment book included notes such as "By taking a pair of steps, I crossed over the threshold of one epoch to another, into the space era." A reporter who visited expressed his interest in signing up for a voyage, noting, "I am going to accompany you on the first flight . . . Please do not refuse my request."[37]

Of more historic significance, in 1931, the Group for the Study of Reactive Motion (GIRD) formed. In 1933, GIRD's space rocketry enthusiasts joined with an army-based weapons group to form the Reactive Scientific-Research Institute, the world's first government-run rocketry institute. Among GIRD's members was Sergei Korolev, an amateur glider builder, with hopes of developing a rocket-powered glider. In the 1950s and 1960s Korolev became known as the "chief designer" of the Soviet Union's space program and mastermind of its triumphs.

• • •

While rocket dreams brimmed over in the Soviet Union and Germany, in the United States Goddard persevered. The same month that the first issue of the early science fiction magazine *Amazing Stories* hit American newsstands, March 1926, Goddard and machinist Henry Sachs used a blowtorch to touch off the first liquid fuel rocket in the snows of Goddard's Aunt Effie's farm (she of the cherry tree) in Massachusetts. They watched flames, heard the roar, and then the motionless rocket, holding about five pounds of fuel that doubled its weight, slowly rose, curved off frantically, hit the snowy field and shot along the ground while Goddard yelled, "I think I'll get the hell out of here."

In 1928, when Buck Rogers was serialized in *Amazing Stories*, illustrations of the balding scientist-inventor Dr. Huer were modeled on Goddard. A comic strip and films followed. While in Europe enthusiasts regarded spaceflight as an unfolding reality, in the United States, space rocketry was relegated to the entertainment realm. The first editor of *Amazing Stories*, T. O'Conor Sloane, a retired professor of physics and mathematics, disparaged the notion of spaceflight. He considered it foolhardy, writing, "What could man do if chilly Mars or cloudy Venus were his destination? The time of the journey might run into years."[38] And, in response to a reader's inquiry about joining a society devoted "to the problem" of interplanetary travel, Sloane wrote, "We do not believe in the possibility of interplanetary travel, but the subject has given many good stories." [39]

It was because of the attitude of figures such as Sloane that Goddard kept a low profile, while continuing to build his rockets and file over two hundred patents. A friendship with Charles Lindbergh in 1929 helped Goddard gain funding from the Guggenheim family, approximately $20,000 a year for nine years. By the late 1930s, now situated in Roswell, New Mexico, Goddard's rockets were lifting as high as 7,500 feet, but still well short of "extreme altitudes."[40]

Rocketry fans in the United States remained an enthusiastic minority. In 1929, Gernsback lost ownership of *Amazing Stories* and started a second magazine, *Wonder Stories*, and hired M.I.T. graduate David Lasser as its editor. After editing dozens of science fiction stories, Lasser began to wonder if the motto of Gernsback's first science fiction magazine "Extravagant Fiction Today—Cold Fact Tomorrow" could be applied to space travel. "I researched the problem. I spent evenings poring through technical journals and scientific books . . . The rocket, I discovered could not only operate where there was no air, but could operate *best* where there was no air."[41]

With likely prodding from Gernsback, Lasser gathered science fiction writers to create the American Interplanetary Society (AIS) in 1930, "to stimulate interest in space exploration." Another member explained that its sole purpose was "the promotion of interplanetary travel."[42] The group met first at the Italian restaurant and speakeasy Nino and Nella's, below the apartment of *Wonder Stories* contributors Edward and Leatrice Pendray. Among the club's efforts in 1931 was the signing of a mutual cooperation agreement with the German rocketry group *Verein für Raumschiffahrt* (VfR). Within a year, the AIS had one hundred members, including the young Robert Heinlein.

In 1931, AIS members built a seven-foot rocket at a cost of $49.40, relying on improvised elements such as a malted milk canister as a coolant lining. A test firing in New Jersey failed, but a second launch, from Staten Island, recorded on newsreels, succeeded, and the rocket climbed to 250 feet.[43] Experiments continued through the decade and, at a bid for respectability, the American Interplanetary Society renamed itself the American Rocketry Society—and later became the American Institute of Aeronautics and Astronautics. When the United States entered World War Two, several members started the rocket company Reaction Motors Inc., which received military funding. (The company built the rocket engines for the X-15 rocket plane that Neil Armstrong and others flew at Edwards Air Force base, starting in 1959.)

By 1933, Lasser became involved in organizing unemployed workers to agitate for relief. A diehard capitalist with no sympathy for socialism, Gernsback fired Lasser. Soon after, Lasser resigned from the American Interplanetary Society and spent the rest of his career as a labor organizer. Lasser's main legacy to the spaceflight idea was his book, *The Conquest of Space* (1931) which sold poorly, but was reprinted in Britain where one of his youthful readers, Arthur C. Clarke, claimed it prompted his interest in science and science fiction.

In *The Conquest of Space*, Lasser expressed hope that a "generation that grew up space-minded from their childhood could create, from that new perspective, a sense of community of mankind, sharing a common destiny on this little earth."[44] Developing an "interplanetary mind" would prompt the formation of a just and unified community on Earth. Lasser was promoting what science fiction fans, in the 1940s referred to as the "long view." Like nineteenth-century French scientists, Lasser believed that scientific advance and education could reform society. Space exploration would encourage a breakdown of racial hierarchies, encourage inclusion, cooperation, and promote peace. Lasser proposed, as Tsiolkovsky had before him—and Gene Roddenberry after—that space exploration become an international effort. He suggested that an "International Interplanetary Commission" be established in Switzerland.

Both Fedorov and Lasser linked space expansion to spiritual growth, and the disappearance of instincts such as nationalism and militarism, as humans learned to be citizens of the cosmos. One of Fedorov's favorite notions was that military cannons, if turned skyward, could encourage rainfall and improve humanity's lot, serving peace. Perhaps the rocket, born of war, could serve progress. Yet as Verne had implied with his Gun Club of Baltimore, spaceship technology would most likely be the product of military contractors and wartime pressures. While Fedorov and Lasser linked the spaceship to the end of warfare, during World War Two Wernher von Braun, a Nazi military officer, ushered into being the V-2, the world's first rocket capable of escaping the Earth's atmosphere.

# CHAPTER THREE:
# VON BRAUN

• • •

"We desire to open the planetary world to mankind."
—Wernher von Braun, "Reminiscences of German Rocketry," 1956

"The principal methods of killing used at Nordhausen Concentration Camp
were shooting, hanging, strangulation, beating, starvation, exposure, deprivation
of medical attention and overwork."—Nordhausen-Dora Concentration Camp
War Crimes Trials Report, April 15, 1948

*I Aim at the Stars,* the 1960 biopic about Wernher von Braun, German
rocket scientist turned Space Age idol, opens with two blond boys
setting off a small rocket from the window of their aristocratic family
estate. It soon shatters the roof of a nearby greenhouse. Whether or
not this scene was apocryphal, young Wernher von Braun clearly was
one of the young converts to the rocket mania that struck Germany in
the 1920s as it also had in the Soviet Union. According to his memoir,
he and his brother attached fireworks to a wagon that they rode as it
careened through the streets of Berlin. They were inspired by the feats
of Austrian inventor Max Valier, who with the funding of automobile
magnate Fritz von Opel, developed a rocket-powered car, a rocket rail
car (the wheels broke off), a rocket ice sled that left the ground, a
rocket bicycle, and a rocket plane that actually flew. Valier had also
written *Der Vorstoss in den Weltenraum (The Advance into Outer
Space)* in 1924, popularizing the more technical and overlooked effort
of mathematician Hermann Oberth.

After poring over a copy of Oberth's 1923 *Die Rakete Zu Den
Planetenraumen (Rocket to Interplanetary Space),* von Braun realized

he would have to study math to pursue rocketry, as Oberth's tract was otherwise "gibberish."[1] In 1929, after graduating high school, von Braun sought out Oberth. This was the same year that Fritz Lang's film *Frau im Mond* (*The Woman in the Moon*) was released, on which Oberth had served as an advisor; for the film's debut, Oberth attempted but failed to create a liquid fuel rocket launch.

Von Braun befriended another young rocketeer, Willy Ley, and then joined Oberth's circle of rocketeers in the *Verein fur Raumsahiffarlit* (VfR, or Society for Spaceship Travel), which was founded in 1927 by Rudolf Nebel, Valier, Ley, and others. By 1930 it had as many as one thousand members, and local officials agreed to let the group test rockets on an abandoned ammunition depot outside Berlin that they named the *Raketenflugplatz* (Rocket Flight Field Berlin). No one was paid. In the Weimar years, unemployment and inflation were high and fads, including nudism, were popular. Skilled workers joined the VfR for the free rent in various bunkers and the meals provided by a local soup kitchen.

The group was so desperate for funding that when they realized the chief engineer of the city of Magdeberg believed that people lived inside a hollow Earth, they convinced him to provide funds to build a passenger rocket which would establish if the horizon curved up or down.[2] In 1931, von Braun, still involved with the VfR, enrolled at the Institute of Technology in Zurich to study mechanical and aircraft engineering. His experiments in those years included placing mice in centrifuges to determine what acceleration would lead to cerebral hemorrhages.[3] Within a year or two the charismatic von Braun, who maintained ties with the VfR, had become the rocket group's de facto leader.

In spring 1932, German army representatives visited the *Raketenflugplatz*. The army men wanted a weapon and one not forbidden by the Treaty of Versailles. Several months later, they arranged for a test firing at Kummersdorf, an army weapons range sixty miles south. The army officers watched the VfR launch a rocket that veered sideways after rising two hundred feet. They were intrigued but not impressed with the amateur group's approach, and insisted changes must be made. Other VfR leaders wanted to stay clear of military control, but von Braun continued to seek army support. As satirist Tom Lehrer later expressed in song, von Braun was "a man whose allegiance/Is ruled by expedience." Of wooing the German Army, von Braun later told the *New Yorker*, "The Nazis weren't yet in power. We felt no moral scruples about the possible future abuse of our brainchild. We were interested solely in exploring outer space. It was simply a question with us of how the golden cow could be milked most successfully."[4]

Responding to von Braun's pleading, other leaders of the VfR "rather unwillingly" gave him permission to join the army's effort. The larger group disbanded in 1934. At first only von Braun and one assistant were moved to the Kummersdorf base, and they managed a rocket engine test-firing in January 1933. By 1934, with Adolph Hitler firmly in power, von Braun's group, growing in recruits, many from the disbanded VfR, had developed two A-2 rockets that reached an altitude of 1.5 miles. These successes, which paralleled those of Goddard, increased their access to the "golden cow." The following year, Willy Ley, who had some Jewish ancestry, fled the country, first to Britain, then the United States—where he became a science fiction writer, journalist, and fervent rocket promoter, proposing in one 1935 article that rockets deliver mail to remote areas of the United States.[5]

Ley and von Braun remained on a friendly basis after the war, and worked together on magazine articles and television scripts for Disney. Ley noted in a 1947 piece about those early years that while von Braun, who was tall, blond, and sturdy, fit the image of the Nazi's "Aryan Nordic" ideal, his colleague had always thought "the Nazis ridiculous."[6] Ley admitted, "I found no reason to regard v.B. as an outspoken anti-Nazi. But just as little, if not even less, did I find him to be a Nazi. In my opinion the man simply wanted to build rockets. Period."[7]

Von Braun underlined his disinterest in the Nazi party line when he wrote that in the 1930s, "To us Hitler was still only a pompous fool with a Charlie Chaplin moustache."[8] Von Braun felt at ease courting the Nazis, perhaps because of his nationalistic streak and opportunism.

In an early assessment of Hitler, von Braun said he appeared to be a "fairly shabby fellow," but later, apparently enthralled, he spoke of Hitler's "astounding intellectual capabilities, the actually hypnotic influence of his personality on his surroundings . . . My impression of him was, here is a new Napoleon, a new colossus."[9]

In 1937, reluctantly, von Braun became a Nazi Party member and his rocket team moved, via his prompting, to Peenemünde, an island on the Baltic. The base eventually had its own concentration camp, as well as ten thousand laborers that included prisoners shipped from Buchenwald.[10] In 1940, at the orders of Gestapo head Heinrich Himmler, von Braun became an SS officer, though he rarely wore the uniform. Still apparently feeling no moral scruples, von Braun enjoyed his power and prestige, and became a well-known man about town, squiring young women on rowboats off the coast, or about Berlin in his motorcar. In a biography far less rosy than many, historian Michael Neufeld argued, "There can be little doubt that he was a loyal, perhaps even mildly enthusiastic subject of Hitler's dictatorship."[11]

By June 1942 the engineers, machinists, and prisoners at Peenemünde created the A-4 (later named the V-2), a liquid-propellant rocket capable of soaring beyond the edges of the atmosphere and traveling at least two hundred miles to deliver its one-ton warhead. (Under separate command, at Peenemünde the German Air Force also built the V-1 flying bomb or "doodlebug.") In 1943 Britain sent a fleet of six hundred bombers to the facility, leveling much of the island, and killing at least seven hundred people, most of them laborers. With Peenemünde devastated, Hitler demanded that weapon production be shifted to underground facilities. Under the authority of the SS, and its commander General Hans Kammler, forced laborers, including concentration camp inmates, turned mine shafts in the Harz Mountains—near the town of Nordhausen—into the huge underground facility Mittelwerk, for the mass production of the A-4, the "doodlebug," and other weapons. Overworked, underfed, neglected, subject to diseases such as typhus, dysentery, and pneumonia, twenty workers died daily in the mines and their corpses were often left piled in the tunnels.

In the struggle between the SS and the German Army to control production at Mittelwerk, von Braun sided with the army. In February 1944, he was summoned to SS headquarters in East Prussia. Himmler reportedly said, "I hope you realize that your A-4 has ceased to be a toy and that the whole German people eagerly await the mystery weapon." He urged von Braun to join his staff to avoid army red tape and to increase support and funding. Von Braun claimed he answered that the A-4 was "rather like a little flower. In order to flourish it needs sunshine, a well-proportioned quantity of fertilizer and a gentle gardener. What I fear you're planning is a big jet of liquid manure! You know that might kill our little flower!"[12]

Himmler smiled, but four months later, the SS imprisoned von Braun and two other members of his team for undermining the war effort. Von Braun, apparently, had been overheard more than once claiming that his interest was in space exploration and not in creating a weapon, as the war was going badly.[13] This was likely true. That von Braun's interest in creating a spaceship overrode his interest in weaponry seems implied in a curt message from von Braun's army higher-up, Walter Dornberger, in 1941. "If the leading employees would pay more attention to what is actually of burning importance to us, rather than wallowing in future dreams, then deadlines might actually be met."[14] Von Braun's savior from SS imprisonment and possible execution was Albert Speer, the Nazi architect turned war production chief, who insisted the V-2 project could not proceed with von Braun imprisoned.

While von Braun did not spend much time at Mittelwerk, instead maintaining offices at Peenemünde, biographer Bob Ward estimates he made fifteen to twenty visits during the final years of the war. Conditions at the underground facility were appalling. 60,000 slave laborers were employed, many from the nearby Mittlebau-Dora concentration camp. The SS beat and hung those who wouldn't cooperate or who were believed to be sabotaging production. Many workers starved or froze to death. At least 20,000 laborers died. Von Braun later said, "The entire environment at Mittelwerk was repulsive . . . I felt ashamed that things like this were possible in Germany, even under a war situation where national survival was at stake." Neufeld described von Braun as "a lesser war criminal."[15]

When the rockets built at Mittelwerk were finally launched, the Nazis were losing the war on all fronts. In June 1944, the A-4, renamed by Joseph Goebbels, the V-2 (or *Vergeltungswaffe*, "vengeance weapon") began to do its damage. It was a desperation effort. Hitler had wanted 30,000 V-2 rockets at his disposal. Somewhat over 5,000 were built and about 3,100 launched. Six out of ten failed on firing—some because of worker sabotage during production. About 1,600 rockets were fired into Antwerp, Belgium, and 1,400 into Britain, mainly targeting London. About 2,500 people were killed in the V-2 attacks and many injured. Von Braun's reaction was mixed. He was dedicated to the ideal of spaceflight. But he was also a loyal German. He later told a British newspaper, "I felt satisfaction. I [had] visited London twice and I love the place. But I loved Berlin, and the British were bombing hell out of it."[16] He and other team members drank champagne when the rockets began blowing up enemy cities.

Meanwhile, knowing that Germany was on the brink of losing the war, Soviet troops were nearing Peenemünde, and that the German command structures had turned chaotic, von Braun actively planned ways to have Americans capture his team. Receiving contradictory orders from Hitler and the SS, he took the SS's advice to head south into Bavaria where they would more likely be captured by U.S. troops. Though under the custody of the SS, von Braun convinced his escorts that it would be safer if his team spread out. He ended up at a mountain ski resort, Haus Ingeburg, enjoying excellent meals, while the Third Reich disintegrated. He moved quickly after his team learned via radio that Germany's leader had committed suicide. "Hitler was dead," he later recounted to the *New Yorker*, "and the hotel service was excellent." Von Braun sent his younger brother, Magnus von Braun, on a bicycle ride down the mountain to find the American troops and surrender.

At the war's close, the United States actively sought out Nazi Germany's top scientists and technicians—hoping to capture them before the

Soviets. In January 1945, Richard Porter, a rocketry expert at General Electric, was commissioned by the military as a "Field Engineer for a secret assignment." He was part of what was initially called "Project Hermes," later Operation Paperclip. Concerning the V-2, the goal, initially, was to take to the U.S. approximately twenty-five men—preferably less—who could help transport, set up, and launch captured V-2 rockets. One memo from early 1945 explained, "Obviously, we want Germans who are willing to assist, as uncooperative persons would be less than useless. We believe that we will need them for a period of approximately six months."[17] The Allies also had to decide who was or wasn't a war criminal. Those who directly oversaw Mittelwerk were prosecuted. Rarely on site, von Braun was spared. His commander, Walter Dornberger, however, served two years in prison after the war (and then joined von Braun in America).

In 1945, in addition to Porter, American physicists Fritz Zwicky and Clark Millikan interrogated von Braun and other Nazi scientists. Von Braun applied his skills of persuasion. Soon a memo was issued ordering "181 Germans and Families to be moved"—presumably to keep them away from the Soviet sphere (this list likely included scientists not involved in the V-2). Army Ordnance recommended that one hundred of the Peenemünde personnel be transferred to the United States. Later that year the U.S. military brought slightly more than one hundred of von Braun's rocketry experts to America, with a six month contract, subject to renewal. Von Braun and many of the others on his team not only outlasted that first contract, but remained permanently in America and became U.S. citizens.

While their transition was fairly smooth, when news of Operation Paperclip leaked to the public in late 1946, some members of the American press offered alarmist accounts of ex-Nazis serving as a fifth column. A *New Republic* article complained of this program hidden behind a "khaki curtain," and noted "except for protests from angry American scientists, the presence of the Germans near the heart of our military machine has caused little comment." The author, Seymour Nagan, argued that the program was a threat to national security, and that many scientists believed the Germans were "not assets, but military and moral liabilities." Nagan quoted a physicist who insisted von Braun's team and others were at best "clever military engineers" but not scientists. No matter how carefully they were "screened," the Gestapo had also screened them more vigorously and determined them to be loyal Nazis. Nagan concluded the Germans were "undistinguished" and "unreliable" dangers to the nation.[18]

Alfred Africano, former president of the U.S. Rocket Society, wrote in a letter to the *New York Herald Tribune*, "British victims of the V-1 and V-2s would have no doubts at all as to the proper treatment of these scientists as war criminals . . . how amused the X-Nazis must be at our apparent credulity."[19] Other articles echoed these sentiments, with titles such as "Send German Scientists Home," "Jobs for German Scientists Opposed," and "Scientists Shocked by U.S. Efforts to Place Nazis in School Jobs Here." On March 24, 1947, the Federation of American Scientists urged the U.S. government to bar the ex-Nazis from jobs in private industry and education—and suggested returning them to Germany "as soon as possible." Yet others replied the safest place for ex-Nazis was inside the military. An editorial in the *El Paso Herald Post* briefly argued "They'll Do Us Less Damage Here."[20]

Reports in the German press varied. Many were neutral or mildly ironic, while some accounts, particularly from East Germany, indicated outrage. One article from 1948 just noted that Operation Paperclip: "employed a precise list of men and women that had in any way to do with Hitler's 'Wonder Weapons.' 'Nominal' Nazis were not excluded from employment."[21] A Berlin newspaper offered a damning satire in 1947, "Enough is Enough," a dialogue between a child and his mother:

"Mum, I would also like . . ."

"What would you like then, my boy?"

"To be a war criminal."

"Are you crazy? What nonsense are you talking about?"

The boy went on to explain:

"But mother, I'm hungry and in Texas it is so beautiful, there are pineapples, oranges, and dates."

"What does that have to do with being a war criminal, dumb boy?"

"When one is a war criminal, then they go to Texas and receive pineapples, oranges, and dates to eat."

When the mother queries the boy further, he insists that 240 German scientists now lived near El Paso, including their boss.

"Who? Wernher von Braun?"

"Yeah, him, the one who invented the V-2. His bride was also taken along. And they will all soon be Americans. They must only make many V-Weapons . . . They are doing very well, and all the people are awfully nice to them. And there are no ruins there, and no food cards, and no Nazi commissions."[22]

A 1946 diatribe in the *Leipziger Volkszeitung* suggested that when it came to developing devilish weapons there was no distinction between Nazi Germany and the post-war West. It opened with: "What did I tell you! The back-street abortionists of the Third Reich befall America, to complete the 'work' to which they were 'appointed' undisturbed. As 'workmanship'—theirs was certainly in demand, and to those who can serve the devil, he opens to them his magical world."

The editorial continued its unflattering portrait of capitalist America and von Braun. "There . . . those who understand rocket building, they make like a maggot in such luck. The old Hitler youth are ready, to serve the Spirit of the new times; no man speaks of treason, because time is money and money is power and power is, what the heart desires. They will be denazified without embarrassing examination and become citizens of Dollar America. The fatherland is wherever the wheat blooms." It concluded, "Did you previously serve the Third Reich as a mass murderer or as an engineer of satanic annihilation instruments, well you have an opportunity to demonstrate through good deeds, how seriously humanity should take you, but we are finally rid of you and wish you luck and rupture on the first trip to the moon."[23]

With evidence of such sentiments in his scrapbook, von Braun recognized that a solid job in public relations was in order. The German rocket team's first stop was Fort Bliss outside El Paso and south of the White Sands Proving Ground in New Mexico. There they assembled and launched as many as 70 V-2s over the next five years. In Texas, perhaps enjoying oranges and an occasional pineapple, von Braun began to attend church and reported a conversion experience. He and others also gave lectures and concerts in local clubs (von Braun was an excellent pianist) and all worked to develop impeccable English skills. The German scientists' contracts were extended to five years and they were allowed to bring over their families. This included von Braun's young German bride, Maria Louise von Quistorp, also from an aristocratic family, and a relative of his mother's.

In 1948, von Braun's rehabilitation furthered when he was named an honorary fellow of the British Interplanetary Society (BIS). The embrace of the BIS may have been easier to gain than that of Britons at large—according to one story, members of the BIS were drinking in a pub in London in 1944 when a V-2 struck the neighborhood. The British rocket enthusiasts raised a toast to the rocket's German engineers. (Von Braun became friends with Arthur C. Clarke, who served twice as the society's chairman.) In 1949, the "nominal Nazis'" exile in Texas ended and von Braun's team was moved to Huntsville, Alabama, and the army's Redstone Arsenal where their job was to develop guided missiles.

By the early 1950s, von Braun had brought over his parents, given dozens of lectures about his dreams for outer space, and been rehabilitated as a prophet of the dawning Space Age. After unsuccessfully trying to sell a science fiction novel about the conquest of Mars, he attended, at Willy Ley's behest, the Hayden Planetarium's First Annual Symposium on Space Travel in New York in 1951. He met the editor of *Collier's* magazine, who asked him to oversee a series of articles about space travel. Prior to publication of the first of these, science fiction artist Chesley Bonestell confided to von Braun, "Our Collier deal sounds as if it would be the most impressive step toward space travel, at least from the publicity standpoint of a serious approach."[24]

The March 22, 1952, *Collier's,* which began the series, had eight space-themed articles, lavishly illustrated by artists Bonestell and Fred Freeman. (Articles from the issue, such as "Man Will Conquer

Space Soon," "Across the Space Frontier," and "Conquest of the Moon," were all eventually collected in 1953, as *Flight into Space*.) In fall 1952, two more issues took up the idea of journeying to the Moon and establishing a Moon base. Others covered "Man's Survival in Space," and "The Baby Space Station." The eighth and final *Collier's* space issue of April 1954 featured journeys to Mars. With Willy Ley, and Bonestell, von Braun developed these articles into two books, *Conquest of the Moon* (1953), and *Exploration of Mars* (1956).

The *Collier's* articles made von Braun a celebrity. For the initial space issue, *Collier's* sent out 2,800 press kits and arranged numerous television and radio interviews for von Braun. Several weeks later, von Braun wrote to the magazine's publicity director, "I am still trying to get my feet back on the ground after those thrilling, stimulating, hectic days in New York. Fan and crack[pot] letters keep on pouring in at a rate of 10 to 20 a day, but the tide is slowly receding."[25] A few years later, after von Braun was featured in Disney's televised episodes about "Man in Space," von Braun turned down a journalist he knew who wanted to do a profile, explaining, "I had so much personal publicity lately that I am wasting my time answering fan mail and sending out monographed photographs. I have found out the price of glory is just too stiff these days."[26]

While von Braun's proselytizing convinced people that a new Space Age was dawning, he also became a target for those who thought his enthusiasm too extreme. A *Time* magazine critique in late 1952 painted von Braun as a concocter of visions that sent "practical rocket men into a cold sweat." Among his sins, apparently, was his tepid dedication to the Nazi cause. "He is the man who lost the war for Hitler . . . He was thinking of space flight, not weapons, when he sold the V-2 to Hitler."[27] A year later, Jonathon Norton Leonard, science editor at *Life* magazine also suggested that "serious" rocket men found von Braun's visionary proposals alarming. Leonard insisted that these scientists, "when they observe the following that has gathered around him of little boys in toy-shop spacesuits and teen-age enthusiasts with space dust in their eyes, they accuse him of leading a children's crusade toward sure disappointment."[28]

Von Braun found the criticism baffling. In a letter to Leonard, he insisted that he was as "practical a missile man as many of the others you have talked to . . . Having been in the development of guided missiles for more than 20 years, I am only too well aware of many of the difficulties and unavoidable setbacks (and even sacrifices of human lives?) that such a program will entail. Maybe I am just a little more aggressive than some of the newcomers in this field."[29] Annoyed at the controversy over the "go slow" versus "jump in" approach to spaceflight, he readily agreed that a commission of top scientists and engineers should be established to decide how to proceed to take rockets from the realm of weaponry to space vehicles. Without such an effort, space would remain out of reach. In 1953, he asked, in a speech, "How far in the future is manned space-flight? Unless we adopt a well-ordered, carefully thought-out program, I say that it is very far in the future—much too far. If we piously fold our hands in our laps while we await some apocryphal revelation of research, we may have to wait a hundred years before the first men circle the earth in a satellite orbit!"[30]

Von Braun had inherited from Hermann Oberth the conviction that a space station would be the ideal first step into space. In *Collier's* first space issue, von Braun predicted that "within the next ten to fifteen years the Earth will have a new companion in the skies, a man-made satellite that can be either the greatest force for peace ever devised, or one of the most terrible weapons of war—depending on who makes and controls it."[31] This giant wheel would rotate to create artificial gravity. Von Braun claimed that such a space station would not only be ideal for astronomical and meteorological observations, but also could keep an enemy under tight surveillance, and, if equipped with nuclear missiles, assure military dominance. As von Braun said in a 1952 speech in Washington before industrial leaders, "If we can get our ground establishment set up and working and establish our artificial satellite with its space-to-ground missiles ready for action, we can stop any opponent cold in his attempt to challenge our fortress in space! The space station can destroy with absolute certainty an enemy space-craft prior to its launching."[32]

Curiously, though risking sounding like a "war-crazed Nazi"— rather like the Dr. Strangelove character modeled after him in Stanley Kubrick's 1964 film—by promoting such a hardline military stance, von Braun better positioned himself as someone concerned foremost with the strategic interests of the United States. Trying to find ways to milk the golden cow, in his presentations to U.S. military officers, he promoted his view that a space satellite, however costly, would offer the ultimate nuclear deterrent, and, by the way, guarantee American dominance over the entire planet.

Yet to an international audience, less keen on Cold War posturing, von Braun could, after calculating the costs of spaceflight, comment, "Let us hope, therefore, that by the time mankind is ready to enter the 'cosmic age,' wars will be a thing of the past, and instead of paying taxes for armament, people will be ready to foot the fuel bill for a voyage to our neighbors in space."[33] At an international forum in 1953, sounding like Verne's fictional Baltimore Gun Club President Barbicane, he frankly acknowledged the problem of funding a coordinated space program when war had ceased. "I would hardly dare to predict whether such a program can be instituted and carried through for year upon year during these piping times of peace . . . that is [a] wholly different question. . . . It is in actuality dependent upon the urgency felt by the Western Powers regarding the necessity of the development, and upon the amount of money and energy these Powers devote to it."[34]

As the go-to guys for space popularizations, von Braun and Willy Ley began to stress a peaceful vision of outer space exploration. In 1955, Walt Disney produced three "Man in Space" television episodes grouped under the "Tomorrowland" theme: "Man in Space," "Man and the Moon" (von Braun believed a space station also should be put in orbit around the Moon), and "Mars and Beyond" (this time via atomic powered spacecraft).[35] The episodes featured animated sequences, and live footage of Disney, von Braun, and Willy Ley, in spiffy double-breasted suits, handling model rockets and space stations and swapping ideas about the new adventure ahead. The shooting script "Rockets and Space" (likely later titled "Man in Space") had von Braun stating, "If we were to start today on an organized and well supported space program, I believe a practical passenger rocket could be built and tested within ten years. Of course it would be foolish to rush headlong . . . without first following a step-by-step research and development program."[36]

In this same period, indicating his moral soundness and Christian devotion, von Braun's short essay "We Need the Power of Prayer," was included in Lawrence M. Brings's 1958 compendium *We Believe in Prayer* ("Statement on the Value of Prayer by over 400 American and World Leaders" of "all religious groups"). The testimony of von Braun, who had converted in 1945 after attending church in Texas, was printed alongside that of President Eisenhower, J. Edgar Hoover, Lyndon B. Johnson, Billy Graham, Estes Kefauver, W. C. Handy, Norman Vincent Peale, Dinah Shore, and Art Linkletter.[37]

Prayer, of course, helped distinguish the United States from the "godless" Communist bloc. But for bolstering national image, prayer could not beat out technological prowess. In 1955, the United States announced its intention to launch the first space satellite as part of the 1957 International Geophysical Year celebration. The year 1957, however, marked two technological setbacks for America, the first was when Ford Motors announced its brand new line of Edsels, and the second when the Soviet Union, on October 4, 1957, launched Sputnik, the Earth's first artificial satellite. An international audience was thrilled. A Soviet press release proclaimed, "The freed and conscious labor of the people of the new socialist society turns even the most daring of man's dreams into a reality."

Sputnik stunned the U.S. government and military and whipped up a media storm. News outlets began promoting stories of a "science gap" and "missile gap"—one result was the passage of the National Defense Education Act, which spread federal funding to universities. Another result, in 1959, during the Kitchen Debate in Moscow, Vice

President Richard Nixon could only brag that the United States was ahead of its Soviet counterpart in the manufacture and distribution of washing machines, refrigerators, and other household goods. The "missile gap" was not entirely illusory. While the United States had plenty of missiles, the Soviet rockets really were more powerful than those of the United States. Sputnik, as small as a beach ball, weighed nearly two hundred pounds, while the planned payload for the U.S. Navy's Vanguard rocket, designed to carry the first U.S. satellite, was only three-and-one-half pounds. The Vanguard, a state of the art three-stage rocket, included miniaturized circuitry and solar cells in its design; it was, however, a fresh design, approved in 1955, and had fourteen flawed test launches including three in the September before the surprise Sputnik launch.

Von Braun was at a meeting in Washington that included the U.S. secretary of defense when news broke of the Soviets' Sputnik triumph. Von Braun told the secretary of defense, "We can get a satellite up in sixty days." His army commander General John Bruce Medaris said, "Make it 90 days, will you." Von Braun then answered, "O.K., ninety it is." But he confided, "I really meant sixty."[38] Von Braun's team received approval four days later. Closer to the ninety day schedule (eighty-four days later), on January 31, 1958, answering numerous prayers from America's leaders, von Braun's four-stage Jupiter-C rocket (already test-launched two years earlier) successfully carried into orbit the Explorer 1 satellite, designed at the Jet Propulsion Laboratory.

The Explorer launch came four months after Sputnik, and eased Americans' panic that the Soviet Union had an overwhelming edge in technology and science. But that argument had some heft. The Soviet Union's Sputnik II, launched November 3, 1957, two months before Explorer, weighed over 1,000 pounds, which was a stunning achievement—as von Braun noted in a speech, "it could have transported a human into space." Of the Jupiter-C that carried Explorer, von Braun was modest, and downplayed the apparent rivalry between the army and navy. In a 1958 speech he noted, "We have always felt at the Army Ballistic Missile agency that the Vanguard vehicle was an advanced design, compared to our own, and we were too much 'space men' in our hearts and too little interested in inter-service problems not to wish the advanced Vanguard missile and its crew a full success."[39]

The launch of Explorer 1 made von Braun a somewhat ambiguous national hero. It was not possible to entirely erase his Nazi past, but with his lofty goals and telegenic bonhomie, he seemed to float above simpler ethical categories. Bolstered with Disney television appearances, von Braun's image also was improved by the 1960 biopic *I Aim at the Stars*, starring Curt Jurgens, which traced Wernher von Braun's career from Nazi Germany to the United States. The movie included as an antagonist a U.S. officer whose wife and child had been killed by a V-2 bomb, who asked von Braun why he hadn't been hanged at Nuremberg. Even this character learned to appreciate von Braun after his team launched Explorer 1. So could we all. Although von Braun had worn a Nazi uniform, he had not been a Nazi at heart—or so the filmmakers intended to establish. A producer wrote to von Braun before the shoot, "As you know, they are anxiously trying to show that you were no Nazi, although you were a member of the Party and built the V-2 for Hitler."[40]

The movie opened to protests in London and Antwerp (two main V-2 targets), as well as in Munich and New York. It did not impress critics or audiences. The acting was mixed and the script, as critics at the time noticed, wooden. A *New York Times* film critic also noted, "Its synthetic brand of hero worship may be annoying and offensive to some." During hype for the movie, comedian Mort Sahl quipped, "I aim for the stars, but sometimes I hit London."

In the movie, one tense scene set at a conference after a failed A-4 test launch establishes von Braun's noble contempt for the Nazi party. One of the assembled comments, "The quality of the steel we're getting just isn't up to the job. It wouldn't surprise me if it isn't thought a bit peculiar that we don't have a closer contact with the heads of the party."

Jurgens, as von Braun, responds, "Look, I am a scientist, I couldn't care less about all this party stuff. Hitler, or the Man on the Moon, it's all the same to me. Come to think of it, I prefer the Man on the Moon. He's nearer my heart."

After further angry words, the true believer stalks out of the room, and an officer, likely Walter Dornberger, von Braun's mentor and fellow rocket enthusiast, says, "We've been a long time without results. At least in the eyes of the High Command. You dream of spaceflight, but they keep reminding me that in the third year of the war they want practical results. They'd settle for a shorter range than the stars. Like London, for instance."

In an answer that mixed naiveté with cold-bloodedness, von Braun's character replied, "But Colonel, a bullet that can reach the stars can also be fired across the room."

"Get me that bullet, Wernher, before they move in on us."

In the early 1960s, when von Braun's reputation had improved, Julius Mader, an East German reporter, began a new campaign to "out" von Braun as a war criminal. Mader's *Secret of Huntsville: The True Career of Rocket Baron Wernher von Braun* (1963) mixed fact with contrivance but also more clearly presented the horrors of the Mittelwerk facility. But most critiques of von Braun devolved to the realm of dark comedy. Stanley Kubrick's 1964 *Dr. Strangelove, Or: How I Learned to Stop Worrying and Love the Bomb* featured an ex-Nazi nuclear weapons strategist for the Pentagon, modeled after von Braun. As played by Peter Sellers, he has a blonde hairpiece, a thick German accent, and occasionally, uncontrollably, breaks into Nazi salutes with his artificial arm while uttering outbursts of allegiance to *der Fuhrer*.

But by then, von Braun had taken one giant step away from the past—he was no longer a military man. During the Sputnik panic, in July 1958, NASA was created. In 1959, von Braun became director of the Marshall Space Flight Center, the new name for the Army Redstone facility. The Marshall Space Flight Center staff led the design of spacecraft and propulsion systems for NASA. And the youthful president John F. Kennedy soon provided them with a mission. Gathering his 4,500 employees together on July 5, 1961, von Braun said, "And now for the future: The president said the other day that we should go to the moon. Well, he needs some help. He needs our help and the help of all other elements of the national aeronautics and space administration, as well as the country as a whole."[41] The Mercury, Gemini, and Apollo missions followed. At NASA Marshall, von Braun shaped the U.S. space program. The Mercury Redstone rocket carried America's first astronaut, Alan B. Shepard into space orbit, and von Braun's team also designed the Saturn rockets relied on for the Apollo missions.

A soft-focus was required to allow von Braun to bask in the glow of the U.S. space programs' ensuing successes. As historian Michael Neufeld has pointed out, von Braun's centrality to spaceflight faded as the astronauts eclipsed him as icons of spaceflight. Michael Chabon's *Moonglow* (2016) a fictionalized family memoir, depicts—with apparently much poetic license—a version of his grandfather as an engineer and O.S.S. agent, obsessed with spaceflight and Moon colonization, who was sent to Germany at the end of World War Two to track down von Braun. His motives are complicated, as von Braun is a man he admires, loathes, and hopes to kill, particularly after he learns how business had been conducted at Mittelwerk. At an anticlimactic meeting at a space society conference many decades later, his grandfather's assessment of von Braun: "Nobody wanted to hear that America's ascent to the Moon had been made with a ladder of bones . . . He [von Braun] was, by any measure, the luckiest Nazi motherfucker who ever lived." The grandfather also extends his assessment to the United States for its wartime and postwar conduct. "In a fundamental way, both proved and exemplified by the spectacular postwar ascent of Wernher von Braun, Nazi Germany had won the war."[42]

Von Braun fielded many congratulations following the Apollo 11 Moon landing, and sent dozens of thank you notes to colleagues, including Richard Porter, the engineer from General Electric who had interrogated him in the final days of the war and supported bringing the V-2 team to America. Von Braun wrote to Porter, "It is a great satisfaction to us to prove that your confidence in us was justified. I just want you to know that you and your contributions were recalled many times, with gratitude and admiration, during the historic flight."[43]

The next logical steps in von Braun's plans for conquering space, after the Moon landing, were building a space station, a Moon base, and mounting a manned mission to Mars, all of which he believed could be achieved by the 1980s. But the Nixon administration was not all in on a costly space program. After losing the ensuing budget war, a disappointed von Braun left NASA in 1972, the final year of the Apollo missions.

Three decades before Apollo 11, in 1938, von Braun and his Peenemünde crew had thrown a party with the theme of the first explorers to land on Mars. Von Braun had dressed as an old man with a white beard. He assumed that he would be seventy when such a landing would take place. Von Braun, the amoral prophet of spaceflight, was alive, if ailing, when the Viking Lander tested the first soil samples on Mars. He died the following year, 1977.

# CHAPTER FOUR.

# MODERN CONQUERORS OF MARS

• • •

"It must not be assumed that legs would be the best answer to the problem of locomotion. On a planet of low gravity a spring-operated pogo stick arrangement might be much more useful."—Arthur C. Clarke, 1952[1]

"Mars settlers will find travel easy across the low hills and flat deserts. There is probably no danger from hostile life forms, animal, vegetable or mineral." —Otto O. Binder, 1961[2]

In July 2018 NASA announced that the goal of terraforming Mars, that is, making it a livable environment for humans, was impossible.[3] I thought this might cast a shadow of gloom on the Mars Society Convention to convene in Pasadena several weeks later. After all, the Mars Society's purpose was "to further the exploration and settlement of the Red Planet." Terraforming Mars is a decades-old idea, and though it originated in science fiction, it is potentially a case of big science at its biggest: rejuvenating a dead planet and creating of it a new Earth. This is a godlike undertaking.

Mars has an atmosphere less than 1 percent that of the Earth and an average surface temperature of -81.4 degrees Fahrenheit. (The surface can get as warm as 68 degrees Fahrenheit at noon. But at the poles temperatures can drop to -243 degrees.) Its gravitational pull is 38 percent that of Earth's. Basketball players could probably jump high enough for their feet to near the hoop.[4] At one time Mars had an atmosphere thick enough to retain liquid water on its surface in lakes and rivers. Its slight atmosphere now is 95 percent $CO_2$. Most terraforming plans for Mars propose restoring the atmosphere by releasing the carbon dioxide locked up in soil and rock to trigger a greenhouse effect to warm the planet. Simple species next would be seeded to introduce oxygen and the beginnings of an ecosystem, and then, give or take, a few centuries, or millennia, or million years, *presto!* a breathable, livable atmosphere for humans and other imports.

Even before it had gained scientific cachet, this mythic enterprise, bringing new life to a dead world—as outlandish as Edgar Rice Burroughs's notion of a giant factory that maintained Mars's atmosphere by harnessing light's "ninth ray"—was one of the major propellants of twentieth-century science fiction. When Martian plant life had not been ruled out, Ray Bradbury, in *Martian Chronicles* (1951), could imagine terraforming as simply restoring Mars's dormant ecology. "There lay the old soil and the plants of it so ancient they had worn themselves out. But what if new forms were introduced?"[5] When a view emerged of Mars as a barren desert, terraforming plans became more grandiose. Arthur C. Clarke's *The Sands of Mars* (1951) features biologists who scatter plant seeds and ignite the larger of the planet's two moons to create a "second sun" to help warm the planet.

Scientists took to the notion of terraforming by the 1970s, when the Mariner probes made clear the Martian planetary surface did not seethe with life. Carl Sagan, for example, proposed terraforming Mars in 1971—among his suggestions, to plant dark vegetation on the polar ice, or find other ways to darken the surface so the planet absorbed more sun energy.[6] In 1974, a NASA study suggested that introducing a species like blue-green algae might create a breathable atmosphere in about 140,000 years, while lichens might take ten times as long. The process might be speeded up, perhaps, by introducing genetically engineered species ideal to the purpose, while, à la Sagan, sprinkling the polar caps with dust or sand to speed melting of its ice.[7]

In 1982 NASA planetary scientist Christopher McKay published an article that stated we could warm up Mars in a century, and followed up with a 1991 article in *Nature* "Making Mars Habitable."[8] James Lovelock, the scientist who, along with Lynn Margules, developed the "Gaia Hypothesis," which proposed that the Earth, a self-regulating system, could be thought of as a large, living organism (in fact one "tough bitch"), also was intrigued with terraforming. He described it in the didactic 1984 novel, *The Greening of Mars*.

Giving terraforming classical literary form, in 1988, poet Frederick Turner published, in 10,000 lines, *Genesis: An Epic Poem,* celebrating, in elegant verse, the basic scenario in which scientists introduce bacteria to a barren Martian ecosystem to revivify Mars. In *Mining the Oort* (1992), Frederik Pohl proposed coaxing comets to crash into the planet to bring water and nitrogen to its atmosphere. But the reigning heavyweight among terraforming Mars novels is Kim Stanley Robinson's trilogy *Red Mars* (1992), *Green Mars* (1993), *Blue Mars* (1996), which describes the process of terraforming taking place over centuries, including the explosion of nuclear bombs beneath the planet's frozen zones, as humans settle and create a thriving civilization on another planet's surface.

While terraforming is part of the long game that the Mars Society has nurtured since its inception in 1998, plans for just getting to Mars also were needed. Early scenarios tended to bold strokes. In 1951, Werner von Braun detailed a plan that would involve "70 men, traveling in ten spaceships from a satellite orbit around Earth into a satellite orbit around Mars." According to this blueprint, in less than a year, 950 launches from the Earth to the satellite station would transport components and propellant to build (and fuel) the ten Martian spaceships. After a 260 day flight to Martian orbit, three winged space boats would descend to the Martian surface, carrying a crew of fifty people. They would remain on Mars for 400 days, making observations, searching for signs of life, water, and resources, and carrying out experiments. Two of those winged boats, with wings removed, would then blast back into Martian orbit and the seven spaceships would then head back to Earth. During the return, probes would be launched toward Venus. The entire mission would take just shy of three years. Von Braun did not put a price tag on the mission, but noted that when humanity was ready for its "cosmic age," perhaps having lost a taste for war, taxes could be used to fund this extravaganza.[9]

Subsequent plans, if scaled back from this mother of all expeditions, also followed the template of a space station from which interplanetary missions could be staged. President George H. W. Bush, in 1990, encouraged a study of a mission to Mars that included the assembly of "Space Station Freedom" (later the International Space Station), which would serve as a spaceport for probes, cargo carriers, and possibly nuclear powered spacecraft to journey to Mars. The full program would cost $450 billion or more over several decades. It was rejected.

In stark contrast, Robert Zubrin, founder of the Mars Society, and David Baker developed their "Mars Direct Plan" at Martin Marietta Corporation in 1990, which detailed a cost efficient mission to Mars. Zubrin and Baker wished to show that initial exploration of Mars could be done on the cheap—that is with "simplicity and robustness"—using the "in situ" (on site) resources of the planet, for example, to produce return fuel for rockets. They proposed sending an unmanned ship or Earth Return Vehicle (ERV) to Mars that would set up a station to extract methane and liquid oxygen and refuel. This would be followed by two ships, one, a habitat, with astronauts, the other a backup ERV. After a one-and-a-half year mission the refueled ERV would bring the astronauts home, while the other vehicle would remain on the planet for the next mission. The final chapters of Zubrin's *The Case for Mars* (1996) offer his verdict on long-term terraforming. "I would say that failure to terraform Mars, constitutes failure to live up to our human nature, and a betrayal of our responsibility as members of the community of life."[10]

As in Nikolai Fedorov's cosmism, at its core, terraforming harbors a sacred myth, that of resurrection and renewal. And we are as much in need of sacred myths as scrappy plans for visiting Mars. The problem with this myth: a 2018 NASA-sponsored study concluded there was not enough $CO_2$ (either frozen or gaseous) locked in the Martian soil to create an atmosphere. Bruce Jakofsky, of Colorado State University, the lead author of the NASA-sponsored paper, argued that to unlock enough $CO_2$ to make terraforming possible, the top ten feet of topsoil of the entire planet would have to be denuded. The press loved it. One of the earliest responses was John Wenz's "There's Not Enough $CO_2$ to Terraform Mars—Sorry to ruin your plans, Elon."[11] A week later, Andrew Coates repeated the refrain, "Sorry Elon Musk, But It's Now Clear That Colonizing Mars Is Unlikely—And a Bad Idea."[12] In *Wired* magazine, Adams Rogers turned to a broader target, "Sorry Nerds: Terraforming Might Not Work on Mars."[13]

*New Scientist's* article was more nuanced but made clear that the potential for terraforming was virtually nil. Leah Crane interviewed Jakofsky who said that, given the shortage of $CO_2$ available, "It's not that terraforming itself isn't possible, it's just that it's not as easy as some people are currently saying. We can't just explode a few nukes over the ice caps." In the same article, McKay, a longtime advocate conceded the point, in a roundabout fashion. He said, "If there's not enough carbon dioxide, terraforming would take thousands of years or more but it's still possible. If there's not enough nitrogen, you need Star Trek. You need warp drive and tractor beams, you need to pull nitrogen from the atmosphere of Jupiter. It becomes science fiction."[14]

When I pushed through the dragon red doors that led from the underground parking garage, passing a mural emblazoned with the likes of Albert Einstein, Jack Parsons (rocketry pioneer and occultist), and physicist Stephen Hawking, and wandered in the convention space where the Mars Society, several hundred strong, was meeting, I was curious what its members thought of the NASA announcement. O.K., terraforming had always seemed to me a bit deranged. But in my worldview "a bit deranged" is quite acceptable. It is good to imagine our species capable of grand works. But I worried for the Mars Society membership. I assumed the report may have had the impact of Moses smashing the stone tablets when he saw his people dancing before the golden calf. (This from the POV of the idolatrists.) I thought I'd witness the shattering of a paradigm, as when Copernicus imposed heliocentrism onto the cosmos, or Darwin posited utter chance at the heart of nature.

But it wasn't like that. As I made the rounds—bearing my red "Mars or Bust" tote bag, I felt somewhat out of my comfort zone, overhearing

comments like "So it would be like an ion drive?" from cheerful guys
with the mathematical formula for escape velocity emblazoned on their
T-shirts. Yet, I doggedly made my main query the impact of the recent
NASA press release pronouncing terraforming impossible. No one was
that eager to discuss it. Ethan Cliffton, an architect whose long career
includes designs for a Mars Base as part of a NASA contract to Martin
Marietta's astronautics group, scoffed. He suggested if you went to
Canada's tundra, and seeded a simple plant, or went to Mars's polar
ice cap, scattered it thoroughly, it would start oxygenation. "Terraforming is like sperm. You need to spread it over a wide area. Give it a
million years." No need for short cuts. He had the long view.

Graham Eyre, a thickset forty-something nurse from Britain, and
a recent recruit to Mars fever ("people attending a long time really
are Mars nuts, but I just think it's an idea whose time has finally
come") also shrugged off the NASA bulletin. "I think it was Isaac
Asimov who said that everything is impossible until somebody does
it." Others had similar reactions, such as, "It was one study. You know
how many studies there have been and will be?" Frank Crossman,
who worked for three decades as a materials engineer at Lockheed,
was a bit more cautious. Human colonization of another planet was
a long-term project—it could be a million years. "We have a lot of
problems to address." For him terraforming was the least worry. Foremost, "there has to be economic opportunity," or such an endeavor
will falter.

In the conference center's commons, a demo table was set up for
the board game "Terraforming Mars." Its creators set the game in the
2400s when "giant corporations, sponsored by the World Government
on Earth, initiate huge projects to raise the temperature, the oxygen
level, and the ocean coverage until the environment is habitable." Representing corporations, players "work together in the terraforming
process, but compete for getting victory points." The game board
included a beguiling map of Mars broken into hexagonal sectors,
where players set down tiles to represent projects as well as glittering
copper and green markers to keep track of progress and setbacks.

I finally spotted Dr. Zubrin, founder of the Mars Society, signing books. In addition to his advocacy of Mars' missions, he held more than forty patents, most having to do with the generation of rocket fuels. (The exception proving the rule: U.S. patent 3,652,091, filed when Zubrin was just shy of twenty, providing plans for a "three-way chessboard.") I approached Zubrin, prepared to ask about terraforming, with the same apprehension I'd have approached evangelist Billy Graham with the question "Just checking—is God dead?" I had studied videos of Zubrin in debate mode and he can be formidable. (When, earlier in the year, scholar Marcie Bianco penned the op-ed: "The patriarchal race to colonize Mars is just another example of male entitlement," Zubrin tweeted in response, "Possibly the stupidest article ever written on the subject of human Mars exploration, but also the most repugnant, as it not only attacks the value of the endeavor, but the right of people to pursue it.") I decided not to ask him about the failings of patriarchy, and instead what he thought of the NASA study rejecting terraforming.

He looked up, said, "First of all it was not a NASA study. It was by a professor. And I think the conclusion was unfounded." He recounted how the lead researcher, Jakofsky, analyzed information gathered by NASA's satellite MAVEN (Mars Atmosphere and Volatile Evolution mission). "He was arguing from data, I'll grant him that." Jakofsky's team compared the presence of heavy to lighter isotopes of argon in Mars's upper atmosphere—and at the planet's surface—and from this proportion estimated how much $CO_2$ likely had been lost. "They put it at a half bar of $CO_2$," Zubrin noted. "But saying how much was lost doesn't accurately tell us how much is left. Evidence that there were large lakes even seas on the surface of the planet implies Mars would originally have had 2 bars of $CO_2$. There is plenty left. The data disputed his conclusion."

According to Zubrin there was reason to believe there was enough $CO_2$ still trapped in the rock to thicken the sickly-thin Martian atmosphere. At least one bar of pressure would be necessary to bring the temperature above 0 degrees centigrade. (Atmospheric pressure at sea level on the Earth is just slightly over one bar.) To nail home his point, Zubrin brought in one of his homely, but effective analogies, "If you told me you lost your wallet with $100 in it yesterday could your net worth be calculated? No. In short, to find out how much $CO_2$ is in the soil they need to drill the soil. And not just the surface. They need to drill 200 meters down." He waited to see if I had any objections, then asked, "Have you read my books?" I told him I had read *The Case for Mars*. Sizing me up as a comedy fan, he said, "Well then you'll like this one, it's funny."

He tapped a copy of *How to Live on Mars: A Trusty Guidebook to Surviving and Thriving on the Red Planet*. It was $10. The bio did look amusing, attributing Zubrin's birth to the year 2071 and indicating he graduated from Heinlein High in 2099. "Due to an unfortunate accident that caused his parents' payoff to the school administration to be misplaced, he was mistakenly ranked near the bottom of his class and was forced to accept employment from NASA for seven years (a time span he calls his dark period)." He signed my copy, "See you on Mars."

Virtually every person at the conference expected that the person most likely to get our species to Mars was Elon Musk, founder of SpaceX. Elon Musk has forged an image as a leader of heroic enterprises, a modern Prometheus straight out of science fiction—rather like Jules Verne's Captain Nemo or Marvel's Tony Stark/Iron Man. In addition to youthful and highly profitable forays into Internet start-ups, developing solar power systems (Solar City), and marketing the hippest electric sports car—and one of the weirdest pickup trucks of all time (Tesla), Musk also has plans to create underground pneumatic transportation systems (i.e., the Boring Company's hyperloop— Chicago was briefly interested), and to settle Mars.

Musk has insisted that his only reason for creating vast industries such as Tesla was to underwrite his project of colonizing Mars. In 2016 Musk set 2022 for SpaceX's first manned mission. (As of 2019, that was readjusted to 2024.) In his grand vision, within a hundred years, a fleet of 1,000 reusable rockets holding 100–200 passengers per flight will make trips to Mars every two years. (These flights would be timed to the oppositions when the planets were closest.) Passengers will pay about $100,000 each and ultimately establish a colony as large as one million people on the Red Planet. Musk has quipped, "I'd like to die on Mars, just not on impact." (Because of the thin atmosphere, parachutes don't sufficiently slow landings. Von Braun's idea of a winged lander also would not have worked well. When the Pathfinder Rover was dropped in 1997 both a parachute and retrorockets slowed its descent and a dozen large airbags sprouted around it to ease impact but it still bounced up nearly fifty feet.)

But the summer of 2018 was not the best of times for Musk or his company Tesla. Lauded for its high-end Roadster and Model-S, the roll-out of the Tesla 3, a moderately-priced mass-produced electric vehicle, had been, in Musk's words "production hell." As of mid-summer, Musk was said to be "sleeping on the factory floor" and in a state of crisis. He "alternated between laughing and sobbing" during an interview with the *New York Times*, talked about his reliance on Ambien sleeping pills, lashed out at critics, and appeared unbalanced. "I've had friends come by," he told the reporter, "who are really concerned." Again, as with the NASA press release deeming terraforming "impossible," I thought Musk's recent troubles might cast a shadow on the Mars Society event. Hardly. Virtually everyone I spoke to noted that manned space exploration, once the monopoly of big governments, had passed into private hands, most notably those of Musk.

While Musk did not attend, SpaceX's Paul Wooster, the company's principal Mars development engineer, tall, thin, wearing blue jeans and a trim blue-gray shirt, who looked like a possible stunt double for Musk, was accorded huge applause. His presentation recapped SpaceX history and included footage of the Tesla convertible launched earlier that year into space from a Falcon Heavy rocket to the strains of a David Bowie anthem. Wooster reiterated SpaceX's plans to begin sending cargo missions to Mars in 2022 and, in 2024 two cargo ships and two crew ships to found a Mars base. He admitted this timetable was "aspirational."

Elon Musk, the man behind this grand plan, was born in 1972 in Pretoria, South Africa. His father, Errol Musk, was a mechanical and electrical engineer. His mother, Maye Musk (née Haldeman), was a fashion model and a dietician. His family noticed something different about Elon, the oldest of three children (he has a younger brother, Kimbal, and sister, Tosca). Not just that he would spend hours reading encyclopedia sets. Frequently, it seems, he would withdraw into a trance-like state where he could visually construct models in his mind and work out design and math problems. (Nikola Tesla, the turn-of-the-century electrical inventor is said to have had a similar faculty.)

Although Elon Musk's father doubted the importance of personal computers, he purchased one for his teenage son, a budding video game fanatic who quickly taught himself to code. Musk created his first game "Blastar" in 1984 and published its code in a local computer magazine, along with his description, "In this game you have to destroy an alien space freighter, which is carrying deadly Hydrogen Bombs and Status Beam Machines." Musk was a classic nerd, a reader of science fiction and comic books, and a know-it-all, albeit one who was interested in having a positive impact on the world. He suspected "Maybe I read too many comics as a kid . . . In the comics, it always seems like they are trying to save the world. It seemed like one should try to make the world a better place."[15]

In high school, a gang made the brainy, awkward Musk their main target, assaulting him frequently—once pushing him down a flight of stairs. They even convinced his best friend to lure him out to another beating. After his parents' divorce, Musk moved in with his father, who, apparently, excelled not only at engineering but also at the psychological torture of his offspring. As Musk said of his years with his father, "It was not a happy childhood. It was like misery."[16] At age seventeen, due for mandatory military service to South Africa's apartheid government, Musk moved to Canada, where his mother's family long had roots. He worked odd jobs, enrolled at Queen's University in Ontario, and later transferred to the University of Pennsylvania— fulfilling his longtime hope of living in America. There he studied physics, as well as business. Already he had concluded that there were three critical areas for economic growth that could also fulfill his plan to improve the world: the Internet, renewable energy, and space.

After finishing at the University of Pennsylvania, Musk began graduate work at Stanford, then dropped out, and, with his younger brother Kimbal, in 1995 began the start-up Zip2, which gave companies an Internet presence, as well as maps and directions—combining features now found in Google Maps and Yelp. Newspapers purchased Zip2 software to begin their own city guides. The Musk brothers eventually sold the company, netting $22 million for Musk, and $15 million for Kimbal. Musk next decided to create the first Internet bank, which he named X.com. His company and another Silicon Valley banking start-up, PayPal, competed, and eventually merged. EBay purchased the combined company, PayPal in 2002 in a $1.5 billion deal that gave Musk $250 million.

It was then that he became obsessed with space. He noted, "I went to the NASA website to find out the schedule" for a manned Mars mission, "and I couldn't find it, which was shocking."[17] He decided to leave Silicon Valley for Los Angeles where much of the aerospace industry was clustered, and he also began to attend Mars Society conferences. In 2001 he gave the society $100,000 to help found the Mars Desert Research Station in Utah, where volunteers run simulated Mars missions. Musk also, briefly, served as the Mars Society Director.

After brainstorming with industry insiders, Musk adopted the plan "Mars Oasis." They would send a small greenhouse to Mars that would scoop up Martian soil, and a live video feed would show the birth of "life on Mars." For this project, Musk's brain trust included Michael Griffin—later to head NASA from 2005 to 2009. Determined to reach Mars on the cheap, Musk, Griffin, and other associates traveled to Russia in 2001 to buy unused ICBM missiles. The Russians demanded $8 million per rocket. Musk turned down the offer, after concluding he could build rockets cheaper and so open a new space market. In 2002, he formed Space Explorations Technologies (SpaceX), based in El Segundo, California (near LAX). Project Oasis was set aside. While satellite launches and International Space Station shuttles were the short game, human settlement of Mars was the long game. Outside Musk's cubicle at SpaceX were two large posters of Mars: one as a barren planet, the second a rendering of a green, terraformed Mars.[18]

As Zubrin noted while giving Musk a "Pioneer Award" at a Mars Society conference in 2012, he admired that Musk's goal was not simply to make money. Musk already knew how to do that. No, designing rockets "that sometimes blow up and leave you feeling very unhappy" was a tough endeavor. "What motivates him," Zubrin added, "is to make something necessary and grand happen."

Musk, with his vaguely South African accent, diluted from years in North America, responded at the award ceremony, "The little window has just cracked open where it's possible for life to extend beyond Earth and it seems sensible to take advantage of that window while it's open." With human colonies elsewhere "the probable lifespan of human civilization and the life of consciousness as we know it is going to be far greater." He added, "The thing that actually gets me the most excited about it—it's the grandest adventure I could possibly imagine . . . We must have inspiring things in the world . . . [so] when you wake up in the morning you're glad to be alive."

SpaceX's first three Falcon 1 rockets failed on launch at a U.S. Army base on Omelek Island in the Marshall Islands. The fourth, in 2008, was a success. Musk told the Mars Society audience, "We were able to reach orbit . . . I was so stressed out I didn't even feel elation, I just felt relief."[19] Soon after, as SpaceX neared bankruptcy, NASA came through with a $1.6 billion contract for twelve supply flights to the International Space Station. As of 2014 SpaceX was averaging at least one launch a month. In 2015, SpaceX became the first company to make its rockets reusable; its first and second stages, after detaching, were guided back to launch pads. In 2017, SpaceX reused a first stage rocket on an unmanned supply run to the International Space Station. To further cut launch costs, SpaceX, as of 2018, had begun to build a spaceport outside Brownsville, Texas.

In early 2018, four months before the Mars Society Conference, SpaceX's Falcon Heavy rockets tested well. The company had a contract to deliver astronauts to the International Space Station and there was still a chance those launches would begin the following year. (Now it appears they will take place in 2020.) Although the largest rocket now in use (with five million pounds of thrust), the Falcon Heavy is not as powerful as the Saturn V used in the Apollo lunar program. But not far off is SpaceX's "Starship" (alternately the BFR, the "Big Falcon Rocket" or "Big Fucking Rocket"), more powerful than the Saturn and likely to be the backbone of the fleet to Mars. SpaceX projects the BFR could include forty cabins, holding two to three passengers in each. (In September 2018, SpaceX announced they had at least one paying passenger for a planned BFR flight that would orbit the Moon in 2023.)

While SpaceX had proven profitable, that summer Tesla was faltering. As of fall 2018 Musk's factory had not managed to churn out the Tesla Model 3—a vehicle for the masses—in quantities promised to investors. Musk also had publically mocked stock analysts, leading to backlash, toked marijuana on a podcast, leading to further backlash, scared off top executives at Tesla, and thanks to dubious announcements on Twitter about taking Tesla private, the Securities Exchange Commission had demanded he step down as chairman.[20] In his mid-forties as of this writing, his reputation was somewhat battered; Musk was less a wunderkind and more a bruised veteran.

So, were space fans worried? Not really. When *Popular Mechanics* ran a series of essays "In Defense of Elon Musk," one, with the title "Because He Is a Superhero," noted, simply: "Bruce Wayne. Elon Musk. Tony Stark. Three men worth billions of dollars who care more about solving important problems than living comfortably, but only one of them is real."[21] Zubrin, always a realist, noted that Musk is a risk taker who "skates very close to the edge of the ice . . . it could happen that he would fall off the ice. But even if that happens at this point he has already won the game . . . He has proven that it is possible for entrepreneurial space to work. Even if he should fail there will be a dozen more SpaceX's that will jump in . . . Either he succeeds and gets us to Mars by 2030 or someone else will do it by 2035. The creative forces have been unleashed."[22] Elon's status is further bolstered by the eerie fact that in Wernher von Braun's 1949 science fiction effort *Project Mars: A Technical Tale*, von Braun gave the leader of the Mars colony, elected for five year terms, the honorary title "Elon."[23] Coincidence?

Musk, who likes to wear an "Occupy Mars" T-shirt, of course, is not the only player in what some are calling "New Space." But he is the most credible of the Mars enthusiasts. While Jeff Bezos of Blue Origin has matched some of SpaceX's efforts—Blue Origin also has developed reusable rockets, and both companies are launching satellites in the hundreds to offer global Internet service—it was not until 2019 that Bezos indicated he shared the grand cosmist vision.

Bezos is a *Star Trek* fanatic who modeled his look after Captain Picard and originally was going to name Amazon "MakeItSo.com." While Bezos's main focus has been on space tourism and satellite launches, he is leaving Mars to true believers. Instead Blue Origin will focus on the Moon, space tourism, and a far future in which orbiting space colonies may be populated by trillions of former Earthlings. Earth, in this vision, will become a vast wilderness park. Another entrepreneur in 2018 itching to colonize Mars, Bas Lansdorp, a former Dutch wind energy executive, had the most dubious of plans: to settle Mars with volunteers. And he would fund the project by selling television rights to the reality show.

Like Musk, Lansdorp got involved when he realized how little progress humanity had made in manned space exploration since the Apollo days. Enthralled by images of the Martian landscape streamed back to Earth from NASA's first rover in 1997, Lansdorp concluded that government space programs, including the European Space Agency, wouldn't ever get him there. In 2011, Lansdorp founded Mars One, with plans to launch twenty-four permanent settlers to Mars by 2024—despite its numerological karma, this date was later pushed back.

Lansdorp's Mars One was based on the premise that "Mars is the stepping stone of the human race on its voyage into the universe."[24] A corollary:

the high costs of Mars missions resulted from planning for two-way trips that required eventually getting astronauts back to the Earth. While Zubrin, Musk, and Goddard (back in 1920) had argued the best option was to manufacture on Mars a capable propellant, such as liquid methane from available $CO_2$, Lansdorp took up Henry David Thoreau's edict to "simplify, simplify." *Who said anyone was coming back?* It would be a one-way trip. No return fuel (or "propellant") needed.

With an online application process, Mars One began crew selection in 2013. As of 2018, they winnowed 80,000 applicants down to a final one hundred. This group, eventually, was to be narrowed yet again to six teams of four settlers. The then-current timetable called for six unmanned cargo launches in 2029, making the outpost "operational" by 2030, with the first manned mission to launch in 2031—and a landing on Mars nine months later in 2032.

Some commentators argued that Mars One must surely have been designed as a reality television show in which an elaborate hoax would be played on the Mars "settlers." The pioneers would be duped into thinking they had traveled to Mars when actually deposited on a remote site on Earth to amuse viewers with their actions, à la Jim Carrey in *The Truman Show*—and reality shows from *Candid Camera* to *Meet Joe Schmo* (2003).

Lansdorp replied, "We're sending the best smart people to Mars, they're not going to believe they're on Mars if they're in Earth gravity."[25] More damning of the project's prospects, an M.I.T. study concluded that, as planned, the settlers would all be dead within sixty-eight days of landing. The M.I.T. group's main critique: a farming project, if successful, would create excess levels of oxygen. No technology yet existed to regulate gas levels, or to gather drinking water from Martian soil. Likewise, as many as sixteen launches would be needed to ready the site—not the six proposed by Mars One. And, spare parts for repairs would be lacking.[26] If Mars One were to become feasible, these problems needed to be resolved.

While Mars One seemed the least likely of Mars settlement plans, that summer it still remained within the realm of the possible. For example, Dr. Gerard 't Hooft, a Nobel prize winning Dutch physicist, was an enthusiast. He noted, "My first reaction was like anyone—'this will never happen.' But now look and listen more closely, this is really something that can be achieved."[27] Zubrin, of the Mars Society, gave Mars One an "E" for effort. He commented to a British reporter, "It's interesting as a thought experiment. I mean, Lansdorp doesn't have $6 billion, but NASA does and in terms of going only one way—hey, we're all on a one-way trip to somewhere! But I think it's utterly fantastical that you'll fund a Mars mission with a reality TV show."[28]

Two of the Mars One finalists, as it turned out, were attending the Mars Society conference in Pasadena. Shortly before the conference ended, I tracked one of them down. Hannah Earnshaw was twenty-seven, British, slender, with close-cropped hair. Originally from Birmingham, Earnshaw, who has a PhD in astronomy, was currently a post-doc researcher in X-ray astronomy at Caltech. She said, "My friends are skeptical it's ever going to happen. And a lot don't get it. But they're happy I'm doing something I'm enthusiastic about."

A longtime science fiction fan, her perspective was changed by reading Kim Stanley Robinson's *Red Mars* (1992). She realized the settling of Mars was not simply a crazy fantasy, but a realizable goal. "Six months later I came across the Mars One website and applied." She agreed that there were some ethical issues with settling Mars—particularly the problem of planetary protection. (According to the United Nations Space Treaty signed in 1967, nations involved in space exploration "shall pursue studies of outer space, including the moon and other celestial bodies, and conduct exploration of them so as to avoid their harmful contamination.")

Earnshaw suggested, "We will have to follow a due diligence checklist." Presumably to limit contamination from settlement. As far as what she would miss on Earth, should the mission proceed, "I suppose the ability to travel, explore, go wherever I want." And for the $10,000 question, a few years ago she answered how her decision was affecting her relationship with her boyfriend, "if I do end up getting through to the final groups, then we're going to go our separate ways and that's something we're both okay with."[29] She also added in that interview that the entire crew would be her "life partners," and the venture could be compared "a little bit to a marriage."

Earnshaw, to no avail, tried to point me to the other Mars 100 finalist in the conference hall. Fortunately, for cosmic reality-television-stars-to-be, the Mars 100 group has a large digital footprint. After leaving the conference, I settled into an episode of a podcast where Kay Radzig, an architect from Reno, already in her fifties, with an exuberant, earthy manner, was interviewed by another Mars 100 finalist, Dianne McGrath of Australia. (Thirty-nine of the finalists were from the Americas, thirty-one from Europe, sixteen from Asia, seven from Africa, and seven from Oceania.)

How did she get involved? Radzig answered, "When I first heard about Mars One, it was, you know, through social media and I'm like *what,* I want to do this because, you know, being the thinker that I am, in a nanosecond I pictured myself in a tin can for seven months, living in a very farty atmosphere with three other people, and, you know, I said, 'I can do that.'"[30]

She recognized that a one-way mission to Mars was not for everybody. "It is unfathomable for most people . . . [but] we've always been explorers and we've always lived in tiny groups. . . . our natural curiosity will never be sated. . . . Yes, we're a cruel and mean species . . . however, we are also a species that can think beyond this planet and beyond this galaxy. . . . collectively we're capable of so much." Like Zubrin and others, she believed avoiding outer space would signal stagnation for the species. She found it reassuring to know that there were at least ninety-nine other people on the Earth "that want to do what I want to do . . . to me that's very comforting . . . I am a passionate human and I love my species . . . I look forward to five hundred years from now when we can look back on this and go, 'jeez, what were we thinking?'"[31]

While the entire project crumbled in early 2019 when Mars One Ventures AG (the money generating arm of Mars One) declared bankruptcy, the one-way trip to Mars dream was still very much alive, if dubious that summer . . . and serious Mars 100 finalists could have done worse than attend the Mars Society conference. The place was packed with Mars lovers, but many, with backgrounds in engineering, were realists eager to tackle the hardships Mars settlers could expect. Radiation was the biggest problem cited. The Earth's magnetosphere shields our planet from 99.9 percent of radiation from the Sun and from this and other galaxies, while Earth's atmosphere, particularly the ozone layer, add additional shielding.[32] Mars not only lacks a substantial atmosphere but a magnetosphere. (The Earth has a solid metal core surrounded by liquid metal, creating convection currents that generate electromagnetism. Mars's core, however, is liquid, and generates no electromagnetic field.) High levels of ultraviolet radiation from the Sun and subatomic particles thrown out as part of the solar wind and emitted in solar flares, as well as galactic cosmic rays bombard the Martian surface. Such radiation can damage tissues, generate cancers, cause birth defects, and, in high enough doses, be lethal.

Another survival issue, even though the movie *The Martian* (2015) made it look easy, would be growing food. Scientists in 2017 determined that the planet surface and dust is toxic—full of chloride compounds or perchlorates, released from the planet's regolith (surface rock) after bombardment by the Sun's unfiltered ultraviolet radiation. These compounds along with hydrogen peroxide and iron oxides in the regolith are poisons to bacteria, and would make the creation of soil difficult.

While settlers would probably be healthy young adults, unlikely to follow Fedorov's program of asceticism and abstinence—they, nonetheless, would have to think twice before "becoming fruitful and multiplying." Pregnant women, nursing mothers, infants, and young children would have to be well-shielded from radiation, living, most likely, underground for years. Children born on Mars, with its weak gravitational pull, would have bones too weak to return to Earth to live unless they spent hours each day on a centrifuge that generated artificial gravity. (Sadly, as for the converse, Earthlings arriving on the planet would not evince superhero powers à la Superman or Edgar

Rice Burroughs's John Carter.) These young weak-boned colonists, unfit to negotiate Earth's gravity, would be stuck forever on the colony and elders would have to weather the fury, and possible violence of resentful adolescents. One speaker convinced the crowd that juvenile delinquency and mental health issues might be rampant in a Martian settlement. On the brighter side: punk rock, death metal, and gangsta rap might thrive.

In truth, besides the bad news about terraforming and soil toxicity, there was little here that science fiction writers had not already contemplated . . . The prevailing attitude was that Mars really was the only choice for a local alternative planet. Although its temperature range was harsh, it could be made survivable. There was water. And the planet rotated every twenty-four hours and thirty-seven minutes—which gives everyone a little extra sleeping time. Although the atmosphere wasn't breathable, at least it was not loaded with sulfuric acid, as was the quite thick atmosphere of Venus.

Yes, some effort has been put into the idea of terraforming not just Mars but also Venus. For decades, in pulp fiction, Venus was thought of as a swampy, hot planet, inhabited by aquatic animals ranging from comely mermaids to finned dinosaurs. In its favor, Venus is similar in size and mass to Earth and within a reasonable range of the Sun. In 1961, Carl Sagan, concerned about reversing Venus's runaway greenhouse effect, suggested bacteria might be developed that could thrive on its atmospheric $CO_2$ and convert it into a pure carbon form such as graphite, and so lower the planet's temperature.

At the time, Sagan was unaware that because of its dense atmosphere, pressure on Venus's surface (93 bars) is about ninety times that on Earth, and the average temperature 880 degrees Fahrenheit. Because of the high pressure and temperatures, any graphite made by Sagan's hypothetical bacteria would simply reignite. Even worse news, the atmosphere includes, in addition to $CO_2$ and water, clouds of sulfuric acid. Just to get this out of the way: everyone agrees that Venus would be really tough to settle. The board game makes it a bit easier: "*Venus. A very deadly world. But it has potential! Up among the*

*corrosive clouds, far away from the scorching hot surface, humans*
*have begun colonizing . . . Your Corporation can make a name for*
*itself by building flying cities, reducing the greenhouse effect, and*
*introducing life."* No one at the conference thought cities floating
above the sulfuric acid laced gases of Venus would be a likely short-
term accomplishment.

Indeed, not everyone was even in a rush to get to Mars, despite
the big banner declaring: "The Mars Society – Humans to Mars in a
Decade." There is a sharp split in the space community between those
who prefer robotic exploration of the solar system, with its lower
risks, lower costs, and useful data gain, and those who want humanity
on other planets ASAP. The NASA planetary scientists who spoke at
the conference, despite representing the "exploration rather than settlement perspective," were accorded great respect by the "Mars or
Bust" crowd. While Zubrin liked to joke about NASA's entrenched
bureaucracy, an audience of about 250 gave rapt attention to Abigail
Fraeman of JPL and Carol Stoker of NASA Ames.

Fraeman, an intense young scientist, obviously in love with her
job, focused on the robotic search for organic molecules on Mars,
and also offered a video clip recreating a shimmering lake that—several billion years ago—existed in Gale Crater on Mars "for at least a
million years." This comment, along with the visuals, drew "oohs"
and "aahs." But when asked her wish list for future Mars research,
Fraeman said she'd like another orbiter such as NASA's Mars Reconnaissance Orbiter, and dozens of unmanned rovers to roam the planet's terrain. To her thinking, Mars, at least for now, should be reserved
for remote science. Carol Stoker, of NASA Ames, made even clearer
her distaste for a potential land rush to Mars. Noting that it was quite
possible that life still existed on the planet, Stoker made the case that
the more humans that landed on Mars the harder it would be to distinguish Martian microbes, if DNA based, from those imported from
Earth or Earthlings. She took the planetary protection article of the
Outer Space Treaty seriously. "Let's face it," she said, "We are big bags
of biology."

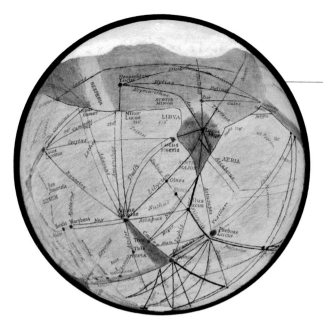

Percival Lowell's 1905 map of Mars highlighting its elaborate canals, which he was certain were the work of highly motivated and intelligent engineers grappling with climate change. *Courtesy Lowell Observatory Archives.*

This *Puck* cartoon of a Martian leaving his airship to urge Earth's plutocrats to stay put indicates that Mars Mania still reigned in 1901. The British play *A Message from Mars*, which the cartoon title referenced, was making its American debut that same year.
*Library of Congress, Prints and Photographs Division.*

Already a celebrated author,
H. G. Wells shipboard circa 1915.
*Library of Congress, Prints and
Photographs Division.*

The elderly Konstantin
Tsiolkovsky, making his rounds
on his bicycle in Kaluga.
*Roscosmos.*

LEFT: Robert Goddard, in Auburn, Massachusetts before the 1926 launch of the world's first liquid fuel rocket from a hilltop on his Aunt Effie's farm (now a golf course). *Courtesy NASA.*

BELOW: Tsiolkovsky imagines weightlessness, from his 1933 *Album of Space Travel. The Archive of the Russian Academy of Science.*

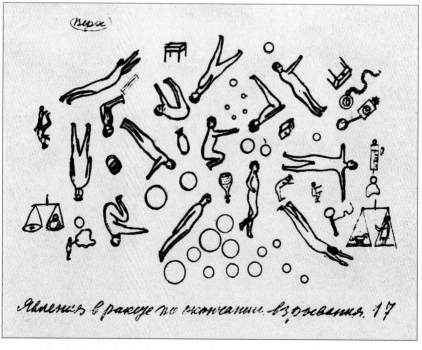

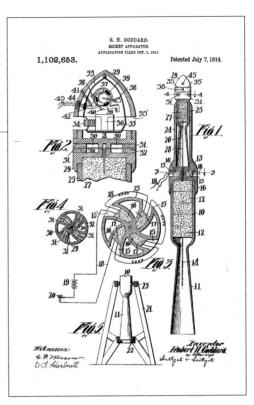

RIGHT: Robert Goddard's 1914 patent for a rocket, equipped with a gyroscope "adapted to transport photographic or other recording instruments to extreme heights" and parachute them back to the ground. He held 214 patents. *U.S. Patent Office.*

BELOW: Nazi V-2 rocket project planner Wernher von Braun (*center*) and younger brother, Magnus (*right*) , at time of surrender to U.S. Army Counterintelligence personnel of the 44th Infantry Division in Ruette, Bavaria on May 2, 1945. *NASA.*

LEFT: Walt Disney with von Braun during planning of Disney's "Man in Space" television specials. *NASA*.

BELOW: Von Braun in front of the Saturn 5 rocket before Apollo 11's launch in 1969. *NASA*.

RIGHT: Elon Musk, in 2014, unveiling the Dragon V2 at SpaceX headquarters in Hawthorne, California. *SpaceX*.

BELOW: Launchpad on Mars as per Elon Musk's vision for a settlement. *SpaceX*.

ABOVE: Mars colonists with view of rockets leaving the Red Planet. *SpaceX.*

LEFT: Gerard O'Neill in the classroom at Princeton University. *Courtesy Tasha O'Neill.*

LEFT: Paintings such as this, depicting the construction of a Bernal Colony at a Lagrange point, helped inspire the space colonies movement of the 1970s. *Artist Don Davis.*

BELOW: The suburban American Dream in an outer space Bernal colony. *Artist Rick Guidice, 1976.*

ACROSS, TOP: Completing construction on a torus-shaped space colony. *Artist Don Davis, 1975.*

ACROSS, BOTTOM: O'Neill colony as a suburban paradise. *Artist Rick Guidice, 1975. All four images © NASA Ames Research Center.*

After evaluating potential designs, O'Neill favored cylindrical colonies as in this interior. *Artist Don Davis, 1975. NASA Ames Research Center.*

In full Buck Rogers garb, underground cartoonist Robert Crumb takes aim at California's Space Day, 1977. *© 2019 by R. Crumb.*

The space yacht *Destiny* – note micro-gravity bubble at center where designer John Spencer would put a hot water sphere as an improvement on the typical yacht's hot tub. *Art by Jeff Courtney. Image courtesy John Spencer.*

Biosphere 2. In this partial view, the dome to the right was one of the two "lungs" that helped stabilize air pressure in the sealed structures. *Photo by Daniel Oberhaus, Wikicommons.*

ABOVE: Hera crew XIX completes their analog mission to the Mars moon Phobos in summer 2019. *NASA*.

LEFT: Richard Branson at Kennedy Space Center in 2006, when pilot Steve Fossett was setting out on Virgin Atlantic's GlobalFlyer for a record-breaking 25,766 mile flight. The aircraft was built by Scaled Composites. *NASA/Kim Shiflett*.

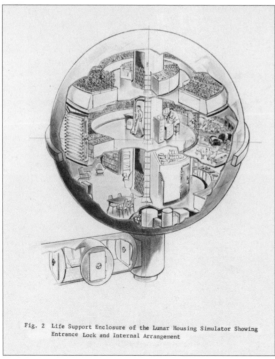

Fig. 2  Life Support Enclosure of the Lunar Housing Simulator Showing
Entrance Lock and Internal Arrangement

ABOVE: Virgin Galactic's WhiteKnight Two "VMS Eve" along with the SS2 for space tourism. *Virgin Galactic.*

LEFT: In 1960 the Martin Company planned a spherical lunar housing simulator complete with farm animals, crops, research, and living facilities. *Courtesy Lockheed-Martin Space Systems Company.*

# Today, Strategic Service Commands are earthbound.

Tomorrow there may be one on the moon.
Because Texaco leads the way.
Today's pilot has the whole network of approximately
600 Strategic Service Command locations available to

him. He knows they're ready when he needs them. With
high quality fuels and lubricants. And the kind of service
which is like having a ground crew he can call his own.
You do the flying . . . leave the service to us.

## Texaco Strategic Service Command.

ABOVE: Texaco announcing in 1970 its potential role on a lunar base. Image courtesy Chevron Corporation. *From Smithsonian Air and Space Museum Archives.*

RIGHT: Bootprint left on the Moon during the Apollo 11 mission. *NASA.*

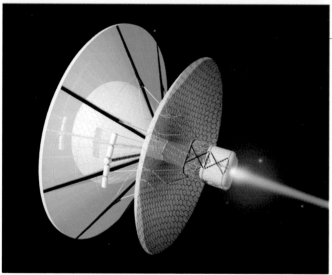

ABOVE: Skidmore, Owings & Merrill's mock-up for the ESA's planned Moon Village—the Earth overhead, offering light in the Moon's nighttime sky. *Image © SOM|Slashcube GmbH.*

LEFT: In 1960, to the delight of science fiction writers, physicist Robert W. Bussard conceived the fusion ramjet starship that can sweep up hydrogen in space to fuel its reactors. The scoop, also, in theory, protected the ship from dust in the interstellar medium. *NASA.*

KEPLER–16b

THE LAND OF TWO SUNS

WHERE YOUR SHADOW ALWAYS HAS COMPANY

Exoplanet Kepler 16b has two suns, offering double sunsets, and is only two hundred light-years from Earth. *NASA/JPL-Caltech.*

NASA/JPL's whimsical Exoplanet Travel Bureau pitch for HD40307g, a very real exoplanet, potentially habitable, with seven times the mass of our home planet, forty-three light-years away. *NASA/JPL-Caltech.*

EXPERIENCE THE GRAVITY OF

HD 40307g A SUPER EARTH

Another doubter: after the conference I heard from science fiction author Kim Stanley Robinson. He, unlike many of the people his books had inspired, was not all that gung-ho on terraforming. His enthusiasm had been dampened by the discovery of the highly toxic perchlorates in the sand and dust of Mars and the potential existence of microbes living below the regolith. He suspected, at least in the short term, the human presence on Mars would be limited to scientific outposts. When I asked about his scaled down vision, he responded, "All along I have considered that the project of terraforming Mars is a great idea that if it ever happens, needs to come in its proper time, which is when we have gotten to a just and sustainable civilization on Earth . . . I regard my Mars trilogy as a good novel, but not a good blueprint."[33]

• • •

While the timetable was vague, NASA had, as of 2017, proposed its own eventual manned Mars mission, Project Deep Space Gateway. Similar to mothballed proposals floated by George H. W. Bush and George W. Bush, Donald Trump signed on to NASA's latest "Moon to Mars" scenario. According to plan, NASA, by 2023, will assemble, a small space station, Gateway, to orbit the Moon and hold visiting human crews of four. Significantly, for Mars fans, by 2029 Deep Space Transport (DST) will be assembled at Gateway. DST will then fly crews on missions to Mars, with a manned mission to Mars orbit set for the 2030s, and a human landing on Mars unspecified, but likely in the 2040s.

Zubrin, predictably, ridiculed Gateway as a waste of tax payers' money—designed more to please contractors with steady work than to complete a mission in an efficient manner. It was probably even worse than the bloated George H. W. Bush plan of 1989. Or as he put it after the conference, "The whole thing is nuts. The Gateway isn't a gateway. It is a tollbooth in space . . . It is not spending money to do something. It is doing something to spend money." As a noncombatant,

I found intriguing at least one aspect of Deep Space Gateway—for the Moon to Mars missions, NASA is investigating a shuttle using solar electric power, i.e., ion propulsion. Its long wing-like solar panels will gather electricity that thrusters will use to turn Xenon gas into plasma and then eject charged ions. Though it will only gradually accelerate, ion engines can reach speeds that exceed those of liquid fuels. Reversing the direction of the engines slow it. One 2009 NASA report suggested that with ion propulsion, the six-month or longer trip to Mars could be cut down to as low as forty days.

Carrying four astronauts at a time on such a shuttle, clearly NASA is not anxious to create a Musk-like colony of a million on the Red Planet. But ion drives? Cool, right? . . . Zubrin's response, "No. It will take longer to get to Mars that way, and require more launch mass for every trip."[34] What can I say? The Mars Society seemed disgusted by the whole decision—mainly because it put off what is already conceivably in reach, i.e., the Mars Direct plan, with SpaceX's Falcon Heavy rockets and the "Starship" (or BFR) in development.

Several weeks after the conference, Fraeman directed me to a colleague, J. R. (John Roma) Skok, a geologist and planetary scientist with SETI, who, like Fraeman and Stoker has been involved in robotic exploration, but unlike them, was very enthusiastic about human exploration of the solar system—his c.v. includes the tag line "Aim for the Stars, End up on Mars."

I asked Skok if, in his opinion as a scientist and explorer, humans really needed to go to Mars? He replied, "Yes. Emphatically. That's the whole point of my entire career." An avid spelunker, as well as a scientist who has led analog Mars missions in Iceland and Antarctica, Skok said, "There are benefits to push people to those limits . . . A future where we are spacefaring is much more interesting than one when we are not. And it is a future worth pushing."[35] Skok noted the biggest obstacles to human expansion into space in the next fifty years were not technological, but political and legal. In the United States, for example, when one administration announces a space initiative, the next tends to cancel it, so it is hard to maintain costly long-term projects. He suspected it would take the founding of an international organization, scientist-led, with funding and clear rules about planetary protection, as well as agreements regulating new industries such as asteroid mining to support a long-term human presence in space.

What did Skok think of private efforts such as those of Musk? "There is skepticism about Musk but seeing his business grow and his successes, as with the Falcon Heavy launch, [these] show he can get things done . . . But he is rather famous for missing his timelines. Having bases on Mars by 2020 is fiction at this point." Whether Mars will one day be home to a thriving colony also is still an open question. "A lot of people see a spacefaring future but not a planetary one. It's possible that people will instead be living on space stations, or on—or inside—hollowed asteroids, without the problem and expenses of overcoming Mars's deep gravity well for landing and takeoff. As a geologist I really do like planets, but is Mars the place to live? Or might it be a tourist site, a place to visit?"

Ultimately, even amongst the truly hopeful, that is, the attendees at the 2018 Mars Society Conference, I found several participants with a similar rather modest hope. Frank Crossman, a retired materials scientist and engineer, who worked at Lockheed for thirty-three years, and Ronald Hattie, an aircraft engineer in Southern California, about forty years younger than Crossman, both said that dreams of colonizing and terraforming were all wonderful but they echoed each other in concluding, "What I would really like is to see a manned mission to Mars before I die."

They just wanted to watch people stepping down on Mars as the Apollo astronauts had in 1969 on the Moon. I agreed that would be cool.

# CHAPTER FIVE

# SPACE COLONIES

· · ·

"An item is listed in the CATALOG if it is deemed: 1) Useful as a tool, 2) Relevant to individual education, 3) High quality or low cost, 4) Easily available by mail." —*The Last Whole Earth Catalog*, 1971

"We want to go to space to save the Earth. I don't like the 'Plan B' idea that we want to go into space so we have a backup planet."—Jeff Bezos, 2016

The last time I saw Taylor Dark III, we were wearing summer yukata (casual Japanese garb), strolling down the humid summer streets of downtown Kyoto passing floats full of glowing lanterns and calligraphy displays to the accompaniment of eerie court music. Our graduate students—one, I recall, was disgusted at my lack of knowledge about the Osaka rock 'n' roll scene—vanished into yet another drinking establishment, leaving us faculty members to celebrate Gion Matsuri on our own . . . Twelve years had passed, Taylor now was teaching government at Cal State Los Angeles, and I was heading south on the Ventura Freeway preparing to connect to the I-210 that led to Taylor's home in Pasadena . . . I also was heading to the heart of the aerospace industry to attend the Mars Society Conference of 2018. ("Maybe I'll join you for a few of the session," Taylor had commented in an email.)

It wasn't until I read historian Patrick McCray's account of science "visioneers" that I understood how Taylor had come upon his interest in space. McCray revealed that my former colleague had a secret past as a teenage space colony fanboy.[1] "It was all very utopian back then," Taylor commented. His interest as a teenager, I suspect, was largely a question of geography. Marooned in suburban Chicago, the issue of space colonies never lodged in my heart or that of my peers—not even the diehard science fiction connoisseurs of the Rutabaga Club. But Taylor grew up in Pasadena, near the NASA/Cal Tech Jet Propulsion Laboratory, making the region one of the main incubators of the aerospace industry (Virgin Galactic also has offices in Pasadena, along with many aerospace contractors). As a citizen of Pasadena, Taylor was a natural to join L-5, the society founded in 1975 to promote life on space colonies.

In the 1970s, a group of space advocates, young and old, decided that the project of settling Mars or even Venus was for wimps. Why escape the gravity of one planet only to succumb to that of another? The century-long obsession with Mars was an example of what Isaac Asimov termed "planetary chauvinism." Far better would be colonies permanently orbiting in zero-gravity. But where? Mathematician Joseph-Louis Lagrange solved this "three-body problem" in 1772, determining the coordinates on a two-body orbital system where a third body could safely "park." One convenient spot was Lagrange Point 5 (L-5) on the Moon's orbit around the Earth, but spaced 60 degrees away from the Moon. Roughly 250,000 miles from the Earth, L-5 was one of the points where a satellite could remain in stable orbit indefinitely. (Unlike the International Space Station, which requires thrusters to sustain its low orbit only two hundred miles above the Earth.) As one L-5 Society bumper sticker, over an image of the Earth, put it in the 1970s: *if you love it . . . leave it.*

Tsiolkovsky first proposed enormous orbiting colonies in the early twentieth century. British scientist John Desmond Bernal also described spheres with ten-mile diameters that could serve as off Earth, gravity-free homes in his 1929 account, *The World, The Flesh, and the Devil.* In the 1970s, in the midst of a dawning awareness of the environmental crisis, Princeton physicist Gerard O'Neill and graduate students cooked up plans for off-Earth colonies that would be rotating cylinders rather than spheres. The idea grabbed the attention of, among others, California governor Jerry Brown, Arizona senators

Barry Goldwater and Morris Udall, as well as L.S.D. advocate Timothy Leary, and science fiction writer Robert Heinlein. The largely student-based L-5 Society, devoted to the O'Neill space colonies cause, developed chapters across the United States and in Europe, including the one that my friend Taylor then had presided over at John Muir High School in Pasadena.

If you wanted to pinpoint the center of the counterculture's excitement over space colonies in the 1970s, a good bet would be Stewart Brand's journal of techno-hippie-optimism *CoEvolution Quarterly*. A former member of Ken Kesey's Merry Pranksters, and the lead organizer of the "Trips Festival" (AKA The Electric Kool-Aid Acid Tests) in San Francisco, Brand has remained, as of this writing, a major broker of West Coast flavored techno-utopianism. In 1965 he began distributing buttons that asked, "Why haven't we seen a photograph of the whole Earth yet?" His hope was that if NASA released such an image, it would stir people to a widened awareness—rather as Flammarion, nearly one hundred years earlier, had urged readers to join him as a "citizen of the skies."

In 1967, NASA released a satellite photograph of the Earth and Brand made it the debut cover of *The Whole Earth Catalog*. With a target audience of young people settling communes with little knowledge of homesteading, the *Whole Earth Catalog* offered an "access to tools." On newsprint and large as a tabloid newspaper, the *Whole Earth Catalog* was a compendium of pithy entries organized around the themes "Whole Systems," "Land Use," "Shelter," "Soft Technology, "Craft," Community," "Nomadics," "Communications," and "Learning." While dedicated to practical skills, it also indicated its origins in the humanistic psychology movement of the 1960s, with its motto: "We are as gods and may as well get good at it."

Browsing, the reader would find entries on wind turbines, breaking in hiking boots (put them on with wool socks, hike across a river, then keep hiking until the boots are dry), wood-burning stoves, "putting up" food, framing houses, fly-tying, solar energy, spot-welding, yoga, tantric sex, surviving a drug bust, and hang-gliding, along with snippets from the thoughts of sixties gurus such as Buckminster Fuller, Gregory Bateson, and Paolo Soleri. While aimed at the counterculture, the *Whole Earth Catalog* made its way into many a suburban house—including my family's home in suburban Chicago. Ten years old in 1967, I enjoyed paging through the *Whole Earth Catalog*. (I was in solid company; Steve Jobs, Apple Computer founder, named it "one of the bibles of my generation.") Our suburban home had been cornfields a decade earlier and the nearby stream was so polluted you wouldn't want to wade in it to break in boots . . . Still, even the catalog's newsprint could make you feel closer to the texture of nature . . . In 1971 over a million copies of the catalog were sold, and the following year, the *Last Whole Earth Catalog* won the National Book Award.

While the general philosophy as well as the chapter heading "Soft Technology" indicate the *Whole Earth Catalog* epitomized the "appropriate technology" and "small is beautiful" ethos animating the back-to-earth movement, Brand was happy to roam the middle ground between technophiles and technophobes. True to the Merry Prankster aesthetic, technology was to be embraced—along with a quiet social revolution. Eventually Brand would go on to champion the emerging technology known as the personal computer (a term he coined), and to cofound one of the earliest online communities, The WELL (The Whole Earth 'Lectronic Link), but first he got hooked on the space colonies idea of Princeton physicist, Gerard O'Neill.

• • •

O'Neill, who grew up reading Tom Swift serials, and listening to episodes of Buck Rogers on the radio, became a radar technician during World War Two and then went on to study physics and become a physics professor at Princeton. He spoke in elegant paragraphs and was stylish: photographs show O'Neill in sports coats and turtlenecks à la Illya Kuryakin of *The Man from U.N.C.L.E.* television series, with a similar choppy hair-do.

Early in his career, O'Neill improved particle accelerators—specifically he designed the storage rings that allow subatomic particles to maintain velocity for hours—but by the late 1960s he was frustrated that other physicists were out-maneuvering him for precious experimental time using those same accelerators. He was tiring of academia. As his wife said of the high energy physics crowd, "They were all fighting for the same dollars. And he was a bit innocent and didn't get the recognition for his groundbreaking work. He was commuting to the Stanford Linear Accelerator Center and secretaries there would type up his notes and many colleagues assumed it all came out of Stanford."[2] He commented, "looking back on it, a certain amount of boredom and frustration was attacking me . . . You had to go through a whole process of writing proposals, defending proposals, either being accepted or turned down . . . always that business of fighting for grants. . . . I got pretty frustrated with it by about the mid-1960s."[3]

In this time of frustration, in 1967, to the amazement of friends, O'Neill applied to NASA to become a scientist-astronaut. Among the finalists, he went through extensive testing. Ultimately, he was not accepted, and suspected his "advanced" age of forty was the reason. NASA's Alan Shepard had contacted him with the bad news, telling him that they "were taking only men who had an estimated better than 50% chance of making a flight in the next four or five years."[4] His disappointment was mirrored in a letter he received from the chemist George Pimentel. Pimentel, who also had applied to NASA, wrote, "This is just a note to tell you of my latest expectations concerning travel to other worlds. I shall confess at the outset that I am in black despair and I am writing to you as much to have a shoulder for a crying platform as to inform you." His friend added that he was "trying to get my personal world reoriented to soften the blow."[5] (Soon after, Pimentel developed the infrared spectrometers used for the Mariner 6 and 7 probes of Mars.)

Unable to enlist with NASA, O'Neill decided to enter the somewhat stigmatized field of outer space studies. He set off in this direction in 1969, when, during an introductory physics course at Princeton he posed the question: "Is a planetary surface the right place for an expanding technological civilization?"[6] At a time of "disenchantment" with big science and technology, O'Neill hoped to show undergraduates that physics could still play a progressive, even revolutionary role. He proposed a bold approach to planet Earth's woes. Why not build a space colony to support ten thousand or more people? Construction costs would be lessened by using materials secured from the Moon. Solar energy might be harvested and beamed by microwaves to the Earth adding value. Likewise colonies could serve as social safety valves and lessen environmental pressures on the Earth. The colonies would offer the beginning of an entirely new space-based civilization. The sputtering millennialism of the sixties had a new outlet.

O'Neill's timing was right. In the sixties and early seventies, critiques of industrial civilization flourished and faith in authority dissipated. According to a Pew Research poll, trust in government to do what is right "just about always" had plummeted between 1964 and 1979 from 77 to 25 percent.[7] Confidence in leaders of the scientific community hovered around 40 percent.[8] While the Apollo moon landings of 1969 stirred excitement, they also brought up questions of skewed spending priorities.

O'Neill launched his Princeton course at the height of the Vietnam War. 1968 had been a year marked by massive student protests throughout Europe, the Americas, and Japan. In 1969, the Tet Offensive turned U.S. public opinion firmly against the war, complete with the nightly news "body counts" that implied U.S. success. In May 1970, during a protest of the Nixon administration's secret bombing of Cambodia, National Guardsmen killed four students at Kent State University. That same year, a faction of the Students for a Democratic Society splintered off as the "Weather Underground," and, embracing anti-imperialist rhetoric, began a campaign that included bombings of as many as twenty-five government buildings.

Beyond social discontent, environmental concerns soured the public's technological optimism. The Santa Barbara oil spill early in 1969 spurred the nation's first Earth Day—as well as the first environmental studies program. The word from think tanks, most notably *Limits to Growth: A Report for the Club of Rome's Project on the Predicament of Mankind* (1972) was that the Earth had finite resources and industrialized society needed to adjust and downsize. Two M.I.T. researchers behind the *Limits to Growth*, Dennis Meadows and Donella Meadows, using computer modeling of "systems dynamics," predicted that human civilization could not maintain its course of exponentially increasing "population, food production, industrialization, pollution, and consumption of nonrenewable natural resources."[9]

If this exponential growth continued there would be a "collapse" or more precisely a "sudden and uncontrollable decline in both population and industrial capacity." To put it more bluntly, by the year 2072—a century from the date of publication—people would be dying off in droves. The dystopian state of affairs might include civil unrest, governmental collapse, famine, global warming, and heightened class warfare. The corporate capitalist mandate to "grow or die" was no longer tenable. At the very least a "steady state" or "no growth" goal had to be encouraged. The authors asked, "Is the future of the world system bound to be growth and then collapse into a dismal, depleted existence? Only if we make the initial assumption that our present way of doing things will not change."[10]

*Limits to Growth* was translated into over thirty languages and sold 21 million copies. As a back cover blurb from a corporate CEO put it, "If this book doesn't blow everybody's mind who can read without moving his lips, then the earth is kaput." Although social conservatives and gung-ho technologists attacked the work, its message sunk in. As economic woes multiplied, heavy industry in the U.S. deteriorated, and gasoline lines grew during the oil crisis, Promethean dreams faded. The era of limits was upon us. Nixon established the Environmental Protection Agency and the U.S. Congress passed the Clean Air Act of 1970, and, to add insult to injury, a universal speed limit of 55 mph was imposed during the final year of the Nixon presidency. Jimmy Carter inherited the mantle of limiter-in-chief. Two weeks into his presidency, during a fireside chat, in which he wore a thick sweater, he stated that "we must face the fact that the energy shortage is permanent." He went on to speak about making "modest sacrifices" and living "thriftily," noting, for example, that people should turn down their household thermostats. A high-calorie nation was being placed on an "energy diet." This did not sit well with rugged individualists or laissez-faire capitalists—or people who liked to drive fast on freeways, i.e., everybody other than VW van owners.

Whatever the merits or flaws of its computer modeling, *Limits to Growth* assumed the Earth was a closed system. This, of course, left a huge opening. Space enthusiasts liked to point out that the Earth was part of a larger "open system"—the surrounding galaxy—with virtually infinite resources. If Earth's resources were, admittedly, limited, why not think big and expand into outer space? O'Neill's space colony idea began to excite attention even though he at first had difficulty publishing his ideas in science journals.

The space colony idea gained momentum in 1972 after O'Neill lectured on the humanization of space at Hampshire College in Massachusetts to a rapt young audience cued in to alternative energy. The lecture had been arranged by Brian O'Leary, then an astronomy teacher at Hampshire. O'Neill and O'Leary had met when both were applying for the astronaut-scientist corps at NASA. (A decade younger than O'Neill, O'Leary was accepted but resigned after training took him to flight school and some gut-wrenching air time. To O'Neill he wrote, "I felt totally out of my element there." He was mocked in the press for saying that "I guess flying just isn't my cup of tea.") After returning to academia, he became an eager cheerleader for O'Neill's bold ideas.

In addition to the undergraduates, O'Neill won over the moderator, a physicist who originally had urged *Science* magazine to turn down O'Neill's article on space colonies. Herb Bernstein, another Hampshire physics professor, recalled that O'Neill's lecture "was well received and treated as science rather than science fiction, though a number of people thought the engineering a lot stronger than the social analysis."[11] Soon after, at a colloquium at Princeton, O'Neill met Freeman Dyson, a physicist at the Institute for Advanced Study with a similar fervor for space exploration and settlement. (Thinking beyond Bernal Spheres, Dyson proposed that advanced civilizations might create massive swarm structures that harnessed the entire energy output of a star.) O'Neill convinced Harold Davis, editor of *Physics Today* to publish his article on space colonies in early 1974.

At the same time, O'Neill sought funding for an academic conference on space colonies. After ten foundations turned him down, O'Neill serendipitously entered the orbit of Stewart Brand. O'Neill applied for a grant at the Portola Institute, dedicated to innovations in education, with one of its projects the *Whole Earth Catalog*. Portola trustee Michael Phillips offered O'Neill $600 for the conference and shrewdly recommended that the money be funneled through Princeton as a formal grant, noting that it would then "generate a lot of red tape. Within many institutions, that is the only reality that is understood."[12]

The university duly publicized the conference, and although it was a small gathering, Walter Sullivan the science reporter for the *New York Times* made it a front page story: "Proposal for Human Colonies in Space Is Hailed by Scientists as Feasible Now." Soon the BBC interviewed O'Neill and he was giving interviews to television and radio stations on both coasts. When the *Physics Today* article appeared in the spring, it hit a nerve. O'Neill claimed only about one in one hundred letters from physicists was hostile. He next published his proposals in *Science* and in *Nature*. With credibility building for space colonies, he received a grant from NASA and planned a second conference for the spring of 1975, on the topic "Space Manufacturing Facilities." Stewart Brand also began to hype the space colonies concept in the pages of his new journal, *CoEvolution Quarterly*.

It helped that O'Neill had scientific standing. Although the space stuff appeared visionary, one year earlier Robert Hofstadter at Stanford had nominated O'Neill for a Nobel Prize for his work on particle accelerators. In July 1975, O'Neill testified before Congress about the feasibility of space colonies. He had learned to stress industrial applications. And, borrowing an idea that originated with engineer Peter Glaser at the Arthur D. Little Company—he posed space colonies as a possible solution to the energy crisis. Glaser, with support from NASA, teamed with researchers from Raytheon and Grumman Aerospace to develop the concept of Satellite Solar Power Stations (SSPS). They proposed three mile long mirror-covered collector panels in Earth orbit that would convert sunlight to microwaves beamed to "rectenna" stations on Earth. JPL scientists tested transmitting electricity via microwaves one mile, and determined a surprisingly high efficiency, suggesting the idea was not entirely utopian.[13] It also was reasonably safe: birds flying through the beam would only become "warm" not cooked.

Soon O'Neill had to balance his academic responsibilities with media exposure that included guest appearances on the Johnny Carson and Merv Griffin television shows, and a PBS interview with Isaac Asimov in which O'Neill assured viewers that both he and his wife would want to live on a space colony. (Tasha O'Neill later told me that at the time this was true. "I would insist that I would open the first restaurant in space, and had to decide which spices I would bring up with me.") In 1976, he warned colleagues that he had done an interview that would soon be published in *Penthouse* magazine. He hoped it wouldn't reflect badly on Princeton. Apparently his colleagues instead basked in the potential notoriety.[14]

O'Neill wrote an account of the space colony idea for general readers which became the 1977 bestseller *High Frontier: Human Colonies in Space*. The book posed itself as a response to the *Limits to Growth*. The Earth was crowded, the environment degraded, and resources depleted. But to aim for a "steady state" was unacceptable. Creativity and diversity would be stifled. What was needed was a new frontier that would allow for innovation, individuality, and freedom. Offering an oddly nostalgic future, O'Neill argued, "What chance for rare, talented individuals to create their own small worlds of home and family, as was so easy a century ago in our America as it expanded into a new frontier?"[15]

O'Neill proposed three general design plans for these "new habitats for humanity:" (1) a giant sphere, perhaps 10–50 kilometers in diameter; (2) a torus—or donut shape, rather like the space station in *2001: A Space Odyssey*; and (3) his own favorite, a giant cylinder, which he called "Island Three." The cylinder's diameter could range from four to fifteen miles, and its length from twenty to seventy-five miles. This upper limit would give the colony the land space equal, roughly, to that of Vermont. All of these colonies would rotate to create artificial gravity and would be "locked in" at Lagrange points 4 or 5—each at a 60-degree angle from the Moon in its orbit.

Spartan at first, O'Neill insisted these cylinders could eventually include sprawling scenery, and a complex ecology, based around three "valleys," where homes would be interspersed with wooded parkland, including features such as an artificial river, bike paths, and mountains (the higher you climb, the easier). Every family would have a garden. O'Neill said in an interview, "There was the question of how to make it as earth-like as possible . . . I had no desire to go the route of just inventing a big spaceship or something that would be a space station. That had no interest for me at all."[16] Colonists could decide whether they would like dense urbanized housing or scattered suburban-style housing. These "new suburbs" would have novel aspects—it might be odd, at first, to see neighboring communities directly above one's home. (An effect that appears toward the end of the 2014 film *Interstellar*.) With industrial wastes "borne away by the solar wind" and no need for internal combustion engines, the habitats would be pollution free. Up to five thousand colonies with populations of ten thousand (i.e., fifty million people) or more could remain at the cislunar (Earth-Moon system) Lagrange points "forever."

To protect from cosmic rays, the entire cylinder would be coated with several meters of Moon rock. Materials from the Moon would be catapulted into space via an electromagnetic "slingshot" or "mass driver"—for which O'Neill had developed a working prototype. (This idea is still popular with Moon mining proponents.) Heavy industry would take place in zero-gravity where materials could easily be manipulated. Farming would be aided with twenty-four-hour light, and pigs, turkeys, chickens, and fish could be raised as additional food.

On this space ark, you could "take along the useful bees while leaving behind wasps and hornets." Entertainment might include low-gravity sports, and swimming under low-gravity conditions. Honeymoon style hotels could also be set up on mountaintops where gravity would be weaker to permit the presumed joys of microgravity sex.

This idea of space colonies enhancing the joys of sex has remained as one of the recurring lures of luxury space tourism. Boosters often quoted science fiction author Arthur C. Clarke: "This much we can predict: Weightlessness will open up novel and hitherto unexpected realms of erotica. About time, too."[17] Neither Tsiolkovsky nor Wernher von Braun published on this idea, but a 1971 issue of the *National Lampoon* featured a black and white photo spread "NASA Sutra" of a nude couple offering poses for space copulation. In addition to terrific places for sex, the colonies could also serve as scientific outposts, manufacturing centers, and solar energy harvesting stations. Some optimists argued that colonists could dehydrate excess food grown in space and drop it out of orbit, to alleviate famine on the Earth.[18]

Ultimately dwellers on these "islands in space," relying on materials harvested on the Moon or asteroids, could shed off new colonies, with some leaving Earth orbit to visit the asteroid belt between Mars and Jupiter. More speculatively, O'Neill suggested splinter colonies eventually could journey to other star systems, "spreading the culture and the species in an expanding sphere from their parent star."[19]

• • •

In the 1970s, as an advisor to California Governor Jerry Brown, Stewart Brand sold the governor on the space colonies concept—a conversion that led journalists and opponents to dub Brown "Governor Moonbeam." Brown declared August 11, 1977 "Space Day." The celebration included speech-making at the Los Angeles science museum and a visit to the Mojave Desert the next morning to watch the space shuttle's maiden flight, a descent from the upper atmosphere after a piggy back ride on a Boeing 747.

At the science museum, Brown, Carl Sagan, O'Neill, NASA astronaut Rusty Schweickart, and various aerospace officials spoke of humanity's future in the stars. While Sagan would eventually support the space settlement idea, on Space Day he discounted it as being as expensive as "one Vietnam War, between one and two hundred billion dollars." During his turn at the microphone, Gerard O'Neill countered that those costs were overblown and insisted robotic exploration had its limits. "Much as we all love Artoo Detoo, I think it's about time we stopped letting him have all the fun. A byproduct of a vigorous thrust into space should be a direct human involvement again." Brown concluded his speech with the note, "It's all there just waiting for you and the rest of the people who stand behind you throughout this world waiting to get into space, go into the oceans, to understand ourselves, and to create the quality of life that our evolutionary potential justifies."[20]

Brand hired underground comic artist Robert Crumb—perhaps best-known then for his character "Mr. Natural" with his tag-line "Keep on Truckin'"—to cover the Space Day event for *CoEvolution Quarterly*. While he initially warmed to the idea, Crumb depicted himself snoozing through aerospace corporations' self-congratulatory sales pitches, and slowly concluded that space exploration would largely be a mask for militarization. (Crumb recently commented, "I remember the Space Day Symposium well. In my comic strip I failed to capture the true level of aerospace corporate aggression that was displayed there, Carl Sagan notwithstanding.")[21]

After remarking "Phooey!" the concluding panels of Crumb's cartoon show him decked out in schlubby Flash Gordon wear, waving a laser gun, and saying, "But what's wrong with space exploration you may ask? Isn't it true that it's an exciting new frontier and that it will raise the consciousness of humanity? Don't be duped by foolish Buck Rogers dreams of glorious adventures among the planets!! Let's wait until we've learned to get along with each other on Earth before we go barging into the cosmos! Whataya say?"

Unlike Crumb, another counterculture icon, Timothy Leary—who in the mid-1960s began to advise kids to "Turn On, Tune In, and Drop

Out," was a genuine enthusiast. After learning of O'Neill's space colonies concept, Leary spun out an entire cosmology called SMI²LE (an acronym for Space Migration, Intelligence Increase, and Life Extension). He began to ruminate on these possibilities while in Folsom prison in 1974 and exchanged letters about human evolution as a possible galactic project with Carl Sagan. (Though skeptical, Sagan wrote to Leary, "I loved your remark about the 'transgalactic gardening club.'")[22]

After his release from prison, in August 1976, Leary helped spread the idea of space colonies with a two day Starseed Seminar for Mutation, Migration, Rejuvenation at the Berkeley Institute for the Study of Consciousness. "Slides of possible L-5 colonies were greeted with delighted 'oohs' and 'ahs' and many excited questions."[23] Of space colonies, Leary said, "I believe it's the destiny of the human species and of planetary life itself to migrate—that we're not terrestrial, we are predestined to leave the womb planet." He also asserted the colonies would allow us to "detach bodies and nervous system from the blind robot direction of the Human Ant Hill."[24] A higher intelligence had "seeded" Earth with DNA (i.e., the galactic project), and we were destined to leave our larval state and expand our minds. With luck, we would merge with a greater Galactic Network.

*CoEvolution Quarterly*'s tome *Space Colonies* (1977) offered a wide range of reactions from an enthusiastic embrace to withering rejection of O'Neill's concept. Many readers believed space colonies, like communes, would allow for new social experiments that could extend groovy lifestyles beyond the planet. In *CoEvolution*'s straw poll, 139 out of 214 readers thought space colonies were a "good idea" and 109 were even ready to immediately "go aboard." Brand's own remarks began, "If built, the fact of Space Colonies will be as momentous as the atomic bomb. Each make statements that are equally fundamental. The one says, 'We can destroy the Earth.' The other says we can leave it, leave home. With that our perspective is suddenly cosmic, our Earth tiny and precious, and our motives properly suspect."[25] Yet the concept clearly fit the Whole Earth mantra "We are as gods and might as well get good at it." No fool, Brand hedged his bet, noting that a purely artificial environment might ultimately drive people insane.

Brand's colleague, Michael Phillips, the Point Institute trustee that donated $600 to Princeton for the first conference was all in. A member of the "International Committee for a New Planet," he believed that by the year 2000 there would be up to five space colonies, with some holding populations of 40,000 or more. Some of these colonies would not orbit the Earth but the Sun. He believed that by 2025, what he called "new planets" would proliferate, and following O'Neill's notion of diversity in the habitats, some might become corporate outposts, while others might be organized around the teachings of Scientology, EST, or Catholicism. More on point, Phillips predicted that by 1986, we would all be connected via a digital "Great Network" and we would "become media freaks."

Phillips's vision of humanity spreading out into a diversity of O'Neill colonies, each conceivably monocultural, has its troubling side—well beyond the frightening thought of having to refuel one's space yacht at a colony permeated with inhabitants spouting ESTian philosophy. Such balkanization in outer space (since realized in cyberspace), as scholar De Witt Douglas Kilgore has argued, would likely result in localized intolerance, expulsions, and encourage a "retreat from public engagement." Kilgore also has insisted that in many "astrofuture" accounts Black Americans appear to be presumed extinct.[26] (Stewart Brand made no effort to hide that space colonies sprang from the white middle class imagination—the inside cover of *Space Colonies* included a vintage Edward S. Curtis photo of two elderly Blackfoot Indians in front of a horse and travois gazing off into the sky—presumably at a space colony—with the word balloon above the elderly man, "Good Bye. Good luck," and the other, a woman, offering the smaller thought bubble, "Good riddance.")

For many of Brand's correspondents, space could wait. Pondering the ethics of space settlement, one reader commented that if and when humans left the planet, "The question is whether we break out of a ruin like a virus exploding from a shattered cell, or leave Earth as a child leaves a playpen."[27] Educator John Holt, a persistent critic of space colonies, referring to the product descriptions found in the *Whole Earth Catalog*, suggested, "Be at least as tough on space colonies as you would a Skil-Saw."[28]

Committed environmentalists were particularly critical. The Earth needed saving. Was it ethical after destroying one's home planet to move on to others? Techno-utopians were wrong-headed when they supported heroic narratives of advancing technology and assumed, as the norm, capitalism's never-ending expansion. Dennis Meadows, one of the authors of *The Limits to Growth* thought we could do without the sizzle of new frontiers. "What are needed to solve these problems on Earth is different values and institutions—a better attitude toward equity, a loss of the growth ethic, and so forth. I would rather work at the root of the problem here." Nature essayist Wendell Berry commented that space colonization represented an outmoded "idea of progress with all its old lust for unrestrained expansion, its totalitarian concentrations of energy and wealth . . . [and] its compulsive salesmanship." And David Brower, president of Friends of the Earth, wrote that he, naturally, could only come up with the knee-jerk reaction expected of an environmentalist, adding, "People who don't have good reflexes are in trouble."[29]

While some critics feared that space development would increase militarization and threats to life and privacy—most questioned the depressing loss of nature in a carefully maintained spaceship. Ken Kesey commented, "A lot of people who want to get into space never got into the earth. It's James Bond. It's a turning away from the juiciness of stuff." William Irwin Thompson wrote, "I don't see anything wrong with setting up a colony in space but I do see something wrong in thinking that one can create wildness by placing it in a container." He added that the "apocalypse that we seek to escape is inside us, and until we come to terms with it, it will follow us wherever—to L-5, to the moon, or to Mars."[30]

Using more down to earth language, Steve Baer, a geodesic dome expert, predicted a degraded life in space colonies. "I don't see the landscape of Carmel by the Sea as Gerard O'Neill suggests. . . . Instead I see acres of airconditioned Greyhound bus interior, glinting, slightly greasy railings, old rivet heads needing paint—I don't hear the surf at Carmel and smell the ocean—I hear piped music and smell chewing gum."[31] George Wald, a Nobel Prize winning biologist feared that O'Neill's interest in "humanizing space" would instead diminish colonists. He wrote, "We are cultivating a race of fractional human beings. . . . the very idea of Space Colonies carries to a logical—and horrifying—conclusion processes of dehumanization and depersonalization that have already gone much too far on the Earth." [32]

• • •

The year 1977 marked the highpoint of the movement. In addition to Brand's *Space Colonies* and O'Neill's *High Frontier*, NASA published *Space Settlements: A Design Study*, based on the 1975 workshop that O'Neill led for NASA Ames and Stanford University. This technical tome was enlivened with full-color illustrations that depicted space colonies as cornucopias packed with trees, houses, gardens, and happy occupants leading apparent lives of leisure in glorified subdivisions of paradise.

That year O'Neill also founded the privately funded Space Studies Institute, which nurtured O'Neill's development of the mass driver, investigated methods of deriving building materials from lunar soil, and explored space solar power, among other projects. O'Neill pointedly insisted the institute was not just another "Gee Whiz" society, but practically-minded—or as he put it, "Hardware is what we do."[33] He was also giving multiple interviews for newspapers, radio, and television talk shows. Although a confident person, his interview notecards (based on his wife's suggestions) included the cues: HOPEFUL—SMILE EVENLY—DON'T BLINK.[34] His public lecture notes painted a semi-utopian image of space settlements: "Not true that advance of civilization must stop and lock into steady-state. Future horizon for our children and grandchildren can be wider, more exciting, with more opportunities . . . Our solar system is a frontier rich in unused resources. It is friendly, and it is waiting for us."[35]

In 1977, the newscast *60 Minutes* did a feature on space colonies. Although generally respectful, the broadcast included a response from Wisconsin Senator William Proxmire, known for his hatred of government waste—including his monthly "Golden Fleece" awards for frivolous government spending. Of the space colonies concept, Proxmire declared, "It's the best argument yet for chopping NASA's funding to the bone. As chairman of the Senate Subcommittee responsible for NASA's appropriations, I say not a penny for this nutty fantasy." (O'Neill had at least one meeting with Proxmire, but according to Tasha O'Neill, "I don't think it went too well. From what I understood the Senator hated the whole idea."[36])

O'Neill understood that academia also had its share of Proxmires. His work, which could be labeled "visionary," bore some stigma. He remarked to one of his scientific colleagues, "Our friend may be right about my work in space development being suspect within the N.A.S. [National Academy of Sciences]; I can see the same response here in Princeton. . . . But I must stand up for my space development work, because the physics of it keeps coming out right. I do feel sure that it will eventually be seen as correct, and if that only happens after I'm gone, well—recognition is nice, but one shouldn't aim one's life toward it."[37] Ultimately he left Princeton for private sector space ventures, including a precursor of the satellite-based GPS system.

Youth support for space colonies, however, was strong. The L-5 society was founded by engineers H. Keith Henson and his wife Carolyn Meinel Henson, in Tucson, Arizona, in 1975—shortly after they participated in O'Neill's second conference on "Space Manufacturing." L-5ers established branches on college and high school campuses across the United States and in Europe. The first issue of *L-5 News*, published in 1975, explained, "The L-5 Society is being formed to educate the public about the benefits of space communities. . . . our clearly stated long range goal will be to disband the society in a mass meeting on L-5." The shorter slogan, available for bumper stickers, "L-5 in '95."

The L-5 Society's newsletters mixed wild enthusiasm with detail-oriented planning, and reflected an ambiguous relationship with the era's environmentalists. With Carolyn editing, *L-5 News* slowly became a sophisticated production. It alternated technical articles on space agriculture, designs for solar collectors, and so on, with interviews with NASA officials, aerospace engineers, and enthusiasts such as Timothy Leary, astronaut Rusty Schweickart—an unofficial liaison between NASA and youth culture—and Brand.

A frequent contributor was T. A. Heppenheimer, a PhD in aerospace engineering and early participant in O'Neill's study sessions. Heppenheimer wrote *Colonies in Space* (published the same year as *The High Frontier*), and his articles mixed an enthusiastic call to adventure with fine-grained engineering details. Heppenheimer admitted that designers would never make a space colony "look like Robert Frost's New England." But he believed satisfaction in work and in relationships could make up for life in a space mall. Settlers would be offered a different kind of nature, namely, "space itself." He added, "There will be new experiments for the human race, new insights, new poetry, and much of this will be strange to us . . . the space colonists will be a people of space."[38]

Like other fan communities, L-5ers had in-jokes and some off-the-wall proposals. For example, when William Proxmire told *60 Minutes* he wouldn't give "one penny" to the "nutty fantasy" of space colonies, L-5 readers dubbed him "Darth Proxmire." One reader wrote, "I hope our grandchildren, living out in the Asteroid Belt or on their way to Tau Ceti, will laugh when they read that statement in their history books."[39] Bruce Friedman's 1977 article "L-5 and the Jewish Community" argued that in a time of crisis space might provide a new refuge, "Jews, of all people, should be interested in space colonization. Space colonization could be considered a form of life insurance if things should get tougher against Jews on Earth. More optimistically, colonization could be viewed as a wonderful opportunity for the expression of Jewish vitality in far-flung and diversified environments."[40] Another reader suggested that since they all believed in "L-5" they should call themselves "elves."

At its peak, with about ten thousand members, L-5 had aspects of a Children's Crusade—if one that never left for the Holy Land. My colleague Taylor Dark, now a professor at California State University Los Angeles, became the president of his newly formed John Muir High School L-5 Society in 1977. A local reporter interviewed the teenagers that year, under the headline "Paradise in heavens sought by space club." The reporter described them as a "far out group," and quoted Taylor saying, "We're very serious, we're working for the idea of a new frontier—a high frontier." The article also quoted a prospectus that vice president Daren Nigsarian wrote for potential recruits, "There is a horrifyingly good chance existence will be misery for you by your 70th birthday . . . The L-5 Society John Muir chapter is working toward a goal that will offer a solution, and make our 70th birthdays better than our 16th."[41]

Corresponding from what the club leaders dubbed "Spacehaven" on campus, Nigsarian reported to *L-5 News* about their group's activities in 1977: they'd sponsored a talk on space colonization by George Koopman, an enthused aerospace promoter and lobbyist, and 750 students had attended. The group planned to raise money at the homecoming fair selling "Spacedust" candy, Star Wars posters, and renting a "Jupiter Jump Spacewalk." Club members also joined letter writing campaigns lobbying for NASA projects, and attended one of the early test landings of the space shuttle at Edwards Air Force base in the California desert. ("We envisioned the space shuttle as the work horse needed to construct the first colony," Taylor commented, as we leafed through his clippings from forty years earlier.)

Nigsarian, looking back, commented that their motives were mixed: "We were fascinated by the promise and potential of space colonization, along with the possible technology and long-term future—but also, in our minds, the club made a "bold statement" of who we were, how smart we were, and how we firmly believed that girls should find those traits irresistible. It would give us standing at John Muir, and a reason, which we really needed, not to hate going to school."[42]

During my visit to Pasadena, Taylor, who still lofts a NASA flag in front of his home, remembered that another bonus for the teenage club members was securing funds to go to an Alcor life extension conference in March 1978 that included guest speaker Timothy Leary proposing that inhabitants of space colonies might live longer lives. Leary "was onstage disco dancing and singing 'staying alive, staying alive,'" Taylor reminisced, "I remember at the end of the meeting—it was near LAX—Leary began running up the down escalator and declaiming. Earlier we had asked him for advice and he said 'you need to present yourselves as an elite. People always want to join an elite group.'"

Nigsarian added, via email, "We were truly absorbed by the righteousness of the mission . . . Only forty solar power satellites in geosync orbit could provide all the energy needs of the United States, and the construction and management of these satellites would be undertaken by a space colony. Raw materials? Easy. A lunar mass-driver, catapulting material off the surface of the Moon. Naturally, everything had a simple answer, which I would rattle-off with the sort of glib certainty that only a teenager can manage."

• • •

The space colonies idea, seemingly within grasp in 1977, never got the funding necessary to move beyond planning. The idea was costly, risky, and untested. After the oil crisis of the late 1970s, decreasing energy prices made the idea less attractive. As Taylor mused in an email, "it all kind of fell apart pretty quickly in the early 1980s, primarily, I think, because it never made engineering or economic sense in the first place. Space Shuttles blowing up didn't help much either. I mean, the original slogan was 'L-5 in 1995.' Once it became clear that it was more likely to be L-5 in 2095, if ever, I think the spirit kind of left the group."[43]

This assessment was clear by 1985, ten years after O'Neill testified before Congress as to the potential bonanza of space colonies, when the American Institute of Physics sponsored the conference "Space Colonization: Technology and the Liberal Arts," retracing the trajectory of the space colony movement. The conference's concluding panel gathered participants from Stanford's 1975 ten-week "Summer Study session of Space Colonization" that NASA had sponsored.

The mood in 1985 was pessimistic. One of the conference organizers, physicist Allan M. Russell, alluding to the enormous costs associated with the project, noted, "Fortunately for us—unfortunately for space colonization—we live in a democratic society. This is a project for a totalitarian society. And it's not going to happen."[44] T. A. Heppenheimer weighed in that the whole notion of solar power from space satellites, considered promising in 1975, clearly was no longer viable. "Economically the power satellite isn't going to cut it, and technically we don't know how to do it. Maybe space tourism, maybe orbiting hotels in lower-Earth orbit are going to be an important opening wedge."[45]

The dialogue soured when audience members began to push an environmentalist and even feminist perspective, using a language of "limits" and "violation." T. A. Heppenheimer lashed out. "I want to talk about this notion of violating space . . . Much of this pessimism, I fear, stems from a view of man as a pitiful, weak creature who must live huddled around his hearth fire never dreaming to look at the stars above. I say NO! All of this stuff about the threat of technology, at least 95 percent of all of it, is nothing more than a put-up job by envious people who resent the attention paid to those who can do rather than merely talk."[46] He went on to rail against people who spoke of the "destructive effects of technology," and concluded, "We have heard a great deal of irresponsible talk in the last twenty years. That is who we technological optimists fight against."[47]

After founding the L-5 Society, Carolyn Meinel became prominent in the hacker movement, and later served as a science and technology advisor at DARPA, and then joined the Institute for Strategic and Innovative Technologies in Austin. When I interviewed her in 2019, she insisted that although they were quite naïve at the onset, L-5 wasn't a silly children's crusade. Their naiveté involved putting faith in the aerospace contractors of the Space Shuttle era and in the idea that as launch costs lowered, space-based solar stations could be assembled.

What were L-5's achievements? She answered that as a community, L-5 brought younger engineers together with the NASA old guard, and inspired space entrepreneurs. An L-5 campaign also helped defeat U.S. ratification of the U.N.'s 1979 Moon Treaty, which, she argued, while idealistic in its efforts to protect resources, "would be a disaster for private enterprise exploitation of space resources and for human colonization of the solar system."[48]

Hope, certainly, springs eternal. Most recently, courtesy of Jeff Bezos, founder of Amazon.com and Blue Origin, and, in his youth, an ardent member of this community. (As his high school valedictorian in 1982, he told a *Miami Herald* reporter that he wanted to build space hotels, amusement parks, yachts, and colonies, to help preserve the Earth.) In 2019, no longer a geeky teenager, but a multibillionaire with a space company, Bezos made a similar announcement. He insisted that space colonies, not Martian settlements, were the true destiny of mankind.

In response, Tasha O'Neill of the Space Studies Institute presented Bezos with a Gerard K. O'Neill Memorial Award for Space Settlement Advocacy. In the months that followed, Bezos announced that the goal was not to create a replacement planet (i.e., a terraformed Mars), but to ease Earth's burden by finding millions of people homes and livelihoods in space. Indeed, trillions of us would eventually be living on orbiting colonies and the Earth would be zoned "light industrial." Blue Origin distributed mock-ups of expansive O'Neill colonies in near Earth orbit stocked with river valleys, green hillsides, waterfalls, a moose, and an eclectic architecture that mixed renaissance European influences with an Americana landscape of barns, grain silos, and town squares. After receiving the award, Bezos noted, "We will have to leave this planet, and we're going to leave it, and it's going to make this planet better."[49]

Interest in space solar power also has revived, with China, India, and Japan all investigating this technology since the beginning of the twenty-first century. Likewise, in 2019, the U.S. Congress budgeted $178 million for space solar power research. Advocates considered this the biggest breakthrough in decades. An official of the National

Space Society, "dedicated to the creation of a spacefaring civiliza-tion"—which formed when the L-5 society merged with Wernher von Braun's National Space Institute—called Congress's decision a "great victory for our cause."[50]

My friend Taylor was less sanguine. He commented, "It seems to me that every five years or so since the 1970s, NASA or the Department of Defense or the National Science Foundation does some new study/research on space solar power, often with the suggestion that it is get-ting more practical. Which it probably is . . . But as far as being cost effective or providing a good 'return on investment' . . . I doubt that is in the cards for many decades to come (if ever)."[51]

Carolyn Meinel, agreed, "I doubt that solar power satellites for Earth-side power will ever make economic sense." But she thought space colonies before this century's end, as per the L-5 and Bezos vision, were still a possibility, if near-Earth asteroids were mined for metal ores, as well as water, carbon, nitrogen, and phosphorus. "So, I see at least one path to space colonies within this century. Whether we will get there requires winding our way past the threats of widespread nuclear war, pandemics, and global organized crime entities fronting as nation states."[52]

When this might be is debatable. At the 1985 conference, Colgate University physicist Charles Holbrow, coeditor of NASA's *Space Set-tlements: A Design Study* (1977) stated, "I do believe that space col-onization in some form is going to happen, because humanity cannot resist the temptation of a new frontier." Another participant, Lawrence Winkler remarked, "it's very funny whenever people come anywhere near the topic, there is an insane, magical feeling about it."[53] Engineer Rowland Richards insisted that space colonies would have to come into being, as "It's still an answer." To what question? "How to build a star ship."[54] Starships, he asserted, massive structures, would have to be built in zero gravity using non-earthly materials. If Richards is to be believed, evolution cannot be denied. We have no choice.

# CHAPTER SIX.

# BIOSPHERE 2

· · ·

—These our witness, Aunty. Us suffer bad. Want justice. We want Thunderdome!
—You know the law: Two men enter, one man leaves.
—*Mad Max Beyond Thunderdome*, 1985

"If they were on Mars . . . They all would have died. If not from $O_2$ depletion,
then starvation."—T. C. Boyle, *The Terranauts* (2016)

In 1967, Lockheed Missiles & Space Company ran a "Five day Lunar Shelter and Extravehicular Manned Test." Through an airlock four employees entered a one thousand cubic foot shelter, housed in a large warehouse. Black and white photographs highlight the crew in baggy jumpsuits doing exercises; sprawled on webbed bunks; changing the plastic insert in the donut-shaped toilet; enduring monotony within steel walls and floors. They look unshaved, sleep-deprived, befuddled. Whatever its value to astronautics, the study was the opposite of an upbeat recruiting pamphlet.

In contrast, recent space simulations promise adventure, whether the desert test-driving of prototype Mars vehicles, astronaut training missions undersea, or excursions from the Hi-SEAS Mars base on the lava slopes of Mauna Loa in Hawai'i. Such simulations usually spring from the loins of institutions such as NASA, ESA, or aerospace corporations. But by far the grandest such simulation was a pure and slightly mad product of the American counterculture. Rather than institutional dreary its creators sought another style—call it late Garden of Eden. Like the optimistic speaker at the Space Colonies ten-year retrospective conference in 1985, they were trying to solve the problem of building a starship.

Their concerns were not with devising antimatter drives or radiation shields, but creating a life support system appropriate for centuries or millennia of dwelling on a starship or new colony. Could a sustainable environment, one that generations of humans would actually want to be confined to, also be portable? Could one establish a travel-worthy version of Earth's greatest ecosystems? And so came the sealed off 3.14 acre geodesic enclosure in Arizona known as Biosphere 2.

In September, 1991, four men and four women, dressed in dark blue jump suits à la *Star Trek* officers, left the larger biosphere to live for two years enclosed in this huge glassed-in structure in the Arizona desert, which included a swathe of rainforest, savannah, desert, marsh, coral reef, and farm. Simulating a colony floating in outer space, the $200 million facility was cut off from the larger biome—more airtight than the International Space Station. The colony, a "techno living synthesis," was not entirely self-sustaining. It depended on sunlight through glassed roofs and walls, and imported electricity to run the sensors, fans, pumps, and computers that helped regulate its systems. But oxygen, food, and water supplies would be self-replenishing, hence the name Biosphere 2. Its creators' goal was to learn how to develop similar biomes on other planets or space colonies.

The project was eccentric, but the media loved it . . . at first. In 1987 *Discovery* magazine proclaimed it "the most exciting scientific project to be undertaken in the U.S. since President Kennedy launched us toward the Moon." But a dozen years later *Time* magazine dubbed it one of the "Fifty Worst Ideas of the Twentieth Century."

The outsized personality behind the project, John Allen, was an adventurer, visionary, member of the Beat generation, and commune founder. Allen was trained as a metallurgist, with a B.S. from the Colorado School of Mines and an MBA from Harvard. He also started an experimental theater group, wrote poetry, plays, and novels under the moniker "Johnny Dolphin," and, with the financial support of Texas billionaire Ed Bass, developed a string of environmental experiments culminating in the grandiose enclosure in Arizona.

Allen was born in 1929 in Oklahoma and at age eleven, already a fan of Tom Swift and Bomba, the Jungle Boy, he read Edgar Rice Burroughs's *A Princess of Mars*. "I dreamed of beautiful Deja Thoris. I would walk to the edge of Grandfather's open porch, stretch out my arms to Mars, and wait faithfully to be transported to that planet."[1] He played football, edited the high school newspaper, at age fifteen ran off to tramp up the west coast picking fruit and working as a lumberjack, and felt a deep attraction to "the whole living, dying, changing, highly differentiated, intricately connected world—the great biosphere of Earth."[2]

After high school he worked as a union organizer in Chicago and served in the Army Corps of Engineers during the Korean War. After gaining his business degree, he went on to work as an administrator of a foreign aid group. One day, he gazed from his office in lower Manhattan at freighters on New York Harbor "and realized I couldn't open my window."[3] Soon after, officially signing on for the bohemian life, Allen boarded a freighter for Tangiers where he would pal around with William S. Burroughs, Brion Gyson, Allen Ginsburg, and Lawrence Durrell. For his life's purpose he decided to follow three lines of action: "theater, enterprise, and ecotechnics," defining this latter term, not so helpfully, as "the ecology of technics and technics of ecology."[4]

In 1967, in San Francisco, Allen and allies started the "Theater of All Possibilities," that they supported with the coffeehouse, "The Sign of the Fool." The theater company, which grew to twenty members, eventually sailed the world doing low-budget tours, while launching adaptations of *Gilgamesh*, and Allen originals such as *Shaxpere and Fitton*, *Brecht and Artaud*, and *Mr. Kabuki*, which began on a New York City subway and ended in outer space. Allen stated the troupe's goal was "to explore man's intentions toward the planet and the galaxy."[5] Jane Poynter (AKA "Harlequin"), one of the first eight "Biospherians," who joined the acting company in London, insisted that they were a "crazy, brilliant bunch," and that "it was a thoroughly bohemian life, and pretty crappy theater—and I loved it."[6]

In 1969, with funding from Allen's wife Marie Harding, or "Flash" (Harding and Allen had married in Vietnam in the early 1960s when she was working as a nurse at an aid center near the Ho Chi Minh Trail), Allen bought a 165-acre ranch in New Mexico that he identified as a "high energy place" and which would become the Synergia Ranch commune. "Synergy," a centuries-old term popularized by inventor Buckminster Fuller, now common in corporate conference rooms, alludes to the notion that a system is greater than the sum of its parts.

Fuller's most famous design, the geodesic dome (or Fuller dome), was inspired by the plans of German engineer and planetarium designer Walther Bauersfeld. Many science historians now consider Fuller as at least equal parts huckster as genius. Allen learned from the master. He attended one of Fuller's hours-long lectures and afterward Fuller asked if Allen would like to see his famous "World Game." Allen said "yes" and Fuller brought out scraps of incomprehensible scribbles. "I took this gnomic behavior," Allen noted, "to mean that the World Game existed in his head."[7] The two tricksters became the best of friends.

At Synergia Ranch, Allen's eclectic lectures and galvanizing presence attracted bright young colleagues. Allen insisted the ranch was governed by Wilhelm Reich's notion of "work democracy." As he put it, you could only critique or speak in areas in which you did some responsible work. Each member split their work days evenly between group projects and individual business projects. Ranch members were charged $45 a month in rent and paid 10 percent of their earnings to cover ranch expenses and improvements. The ranch's enterprises read like those of other communes: gardening, pottery, metalwork, leatherwork, woodwork, automobile repair (for commune members only), as well as theater. Ecological restoration projects became its members' first efforts in "ecotechnics."

On Thursdays, Allen held the floor with metaphysical lectures. He apparently used all his theatrical skills, as he threaded ideas from world religions, philosophy, architecture, sociology, and science into his own speculations, as for example, describing a future undersea civilization.[8] He lectured on Buddhism, Sufism, and esoteric beliefs, and was indebted to the thought of J. G. Bennett, a disciple of Armenian mystic George Gurdjieff. The Synergians were steeped in Gurdjieff's principles that people must develop their skills at feeling, thought, and action to become truly awakened. To achieve this self-liberation was "the work." At the ranch, lectures and study allowed the Synergians to develop their thinking capacity, theater their emotional capacity, and their many enterprises provided a focus for action.[9]

Synergia Ranch disproved the idea that the New Age movement simply involved a taste for crystals, channeling, and uncovering past lives. Then quite vibrant, the New Age drew from occultist systems, but also reflected faith in "transcendence" or shifts to the sacred, and, even more clearly in the case of Synergia, dedication to the concept of "unity" or "holism;" and, undoubtedly, it was grounded in millennialism—the old order was crumbling and an age grounded in new values was dawning.[10]

While Allen sought transcendence—this was a guy who spoke of having had shamanic experiences and running for hours at night, barefoot, chasing coyotes in the desert—his interest was keyed to unity. Like Stewart Brand, he saw no need to reject technological tools or scientific knowledge. Allen's heroes included biologist E. O. Wilson, a student of Konrad Lorenz, and he put great store in what Wilson called the "naturalist trance," a meditative state that allowed one to intuitively grasp details of a larger system.[11]

Like many other New Age groups, the Synergians connected older religious notions of unity to environmental awareness and the development of a "planetary consciousness." A visitor to Synergia Ranch paraphrased one of Allen's speeches as, "Western Civilization isn't simply dying. It's dead. We are probing its ruins to take whatever is useful for the building of a new civilization. This new civilization will be planetary."[12] The Synergians believed that with Biosphere 2 they would not only be reinventing the world but gaining intuitive knowledge of the larger biosphere—knowledge that might save the Earth or the human race.

While many of the era's communes folded, like others that survived Synergia Ranch had a charismatic leader and money-making enterprises. One of their most successful projects, Synopco, the Synergians' construction and design company, built over thirty stucco homes in Santa Fe. In 1973 Synopco was contracted to complete the elaborate Santa Fe home of Texas billionaire Ed Bass, heir to oil and Disney money. Two years later, Bass moved into the Synergia Ranch complex and became a member of the theater company.

Allen, who borrowed the Sufi phrase "create and run," had, a year earlier, in 1972, reported his desire to build a ship and sail the seas. To understand the planet, the Synergians needed to become "sea people"—or as Allen explained, "I had to become a flexible being that directly experienced Planet Water or I would remain in personal unreality about the biosphere."[13] In 1974, Allen and an enthusiastic crew set out from New Mexico to a dry dock in Oakland. They sought expert help on ship-building, and for financial support opened a new café "The Junkman's Palace" on Telegraph Avenue that offered the "greatest hamburger in the world." Over a two-year period, and on a budget of $90,000, the Synergians' Institute of Ecotechnics built the 84-foot-long ship *Heraclitus*.

Modeled after a Chinese junk, with a hull of ferro-concrete, sails, an engine, cabins that could hold a crew of fifteen, and eyes painted on its bow, the *Heraclitus* proved seaworthy, and aided the Synergians in the following decades as they established environmental projects and undertook biological collecting missions in Antarctica, Australia, the Amazon, and the Caribbean. The crew often gave theatrical performances upon anchoring. Although repaired and patched over the years, *Heraclitus* was still making expeditions as of 2017 more than forty years later. Allen and his team may have been "crazy" but they were also gifted.

Continuing with their "create and run" strategy, and aided by the fortune of Ed Bass, over the next decade-and-a-half the Synergians built a hotel/conference center in Tibet, opened an art gallery in London, a conference center in rural France, began two large-scale ranching projects in Australia, undertook the ecological restoration of a coffee plantation in Puerto Rico, and started a large nightclub/arts center in downtown Fort Worth, Texas. The Synergians' Institute of Ecotechnics also hosted yearly symposiums, including conferences on the ecology of the upper Rio Grande in 1973, an Oceans Conference in France in 1974, a Mountain Conference in Nepal in 1978, a Jungle Conference in Malaysia in 1979; a Grasslands conference in 1980; a Planet Earth Conference in 1980, and a Galactic Conference in 1982.

The Galactic Conference, held in France, with luminaries such as Richard Dawkins, Albert Hoffman, the chemist who first synthesized L.S.D., and jazz musician Ornette Coleman in attendance, helped launch Biosphere 2. Ecotechnics Institute member and *Heraclitus's* captain Phil Hawes suggested they could build a spherical spaceship or space colony made of adobe to shield it from radiation. (When Columbia University took over the Biosphere 2 facility in the 1990s, students liked to wear T-shirts emblazoned with the slogan, "My other car is an adobe spaceship."[14]) At the conference, Hawes held up a model of the proposed 110-foot diameter (tiny by O'Neill standards) adobe space colony. It would rotate to create artificial gravity, and at its center, floating theatrical productions would be mounted offering true theater in the round. According to Jane Poynter, after Hawes's presentation, Buckminster Fuller, in an apparent endorsement, said, "I didn't think you could build a spaceship that made sense, but what you've done here does make sense. If you guys don't build a biosphere, who will?"[15]

The idea for the biosphere was inspired by O'Neill's space colonies and science fiction. Likely highly influential was the 1972 science fiction movie *Silent Running* which featured Bruce Dern as an ecologist who maintains greenhouses in space after all of Earth's plant life have been wiped out. When ordered to jettison the greenhouses and take cargo instead, he goes on the run. The theme is purely apocalyptic— Dern, a future-age Noah, becomes caretaker for all the former flora of planet Earth.

Not only science fiction dreams of space settlement drove the project. While offering a blueprint for future outposts on Mars, Allen suggested that on Earth artificial biospheres could provide "refugia" for social and intellectual elites to survive nuclear devastation. In a 1985 pamphlet, Allen wrote, "A hundred Refugia protected by their own energy resources in mountain caverns could release full-scale life . . . after the skies begin to clear . . . Perhaps even the existence of the Refugia . . . could bring home to people and states the gigantic risks they run and thus alter the behavior itself."[16]

But Allen's greater allegiance was toward space settlement. In a 1985 interview, he expressed interest in creating a settlement on Mars, within view of Olympus Mons (nearly fourteen miles high) and near Valles Marineris, Mars's more grandiose version of the Grand Canyon, "the most scenic place in the universe, as far as we now know it."[17] (Unfortunately because of the curvature of Mars, and, paradoxically, the massive size of the mountain, there is no point along the rim of Valles Marineris from which Mons Olympus could be seen. In fact even at the base of Mons Olympus a viewer could not "see it" but only a gently sloping incline to the horizon.[18])

The bold idea of sealing off an entire enclosure from the Earth's atmosphere was entirely reasonable for those who desired to make humanity a spacefaring civilization. And Allen, a disciple of the Russian cosmists, did so desire. He wrote in 1989 that "Biospherics opens up, together with astronautics, the ecotechnical possibilities, even the historic imperative, to expand Earthlife into the solar system and beyond that to the stars and then in time's good opportunity to the galaxies, perhaps in association with biospheres from other origins."[19] In one pamphlet, Allen and coauthor Mark Nelson concluded that while the "fate" of humanity was to be "lost in time, separated in space, entropic in energy and information, limited in matter, deficient in data, uncertain in perception, and certain only of continuing folly," humanity's true destiny, in cosmist fashion, was the "conquest of time, the liberation of space, redundancy of energy and information. . . . the attaining of wisdom."[20]

In his memoirs, Allen, who studied geology and metallurgy, frequently pays tribute to Vladimir Ivanovich Vernadsky, a contemporary of Tsiolkovsky's, and another of Fedorov's students in Moscow. In Russia, few scientists have a higher reputation, in part because of Vernadsky's resistance to both tsarist and Soviet party lines. A geochemist, Vernadsky was one of the first to conceptualize the planet's surface and atmosphere as a biosphere, an interconnected web of organisms and geo-chemical systems. The biosphere, or "living matter" he argued, was being reshaped into a realm of mind, or "thinking matter" that he and his colleague at the Sorbonne, Pierre Teilhard de Chardin, dubbed the "noosphere." Vernadsky offered a holistic view of the intertwining of life and geology. He also offered a saner take on Fedorov's resurrection theme, observing that the living becomes the inert which once again can become the living.[21]

Allen shared the cosmist notion that humans could guide evolution. He reasoned that biospheric systems, "the primary units, the space seeds,"[22] would be necessary to spread human life to space colonies and other planets. He also believed this was a realm where a group like his could make progress. If NASA was not going to take the time, the Synergians could attempt to "build a prototype biosphere for Mars exploration."[23]

NASA had dabbled in the study of artificial ecosystems in the early 1960s, but when it became apparent that people could not long stomach an algae diet, they slowed research for over a decade. Influenced by cosmists such as Tsiolkovsky and Vernadsky, the Soviets persisted. As with NASA, the Soviets, in Bios-1, first had experimental subjects live in a twelve cubic meter chamber with algae as the primary source for oxygen and food. Six weeks was all test subjects could stand. After the subsequent Bios-2, begun in 1968, the Soviet scientists gave up on algae altogether. Then, in 1972, the Institute of Biophysics in Krasnoyarsk opened Bios-3, a sealed stainless steel underground enclosure of 315 cubic meters. It had three grow rooms for crops, a small kitchen, a lavatory, and three small cubicles as private sleeping spaces.

In the hermetic atmosphere of Bios-3, a one million dollar facility, scientists and physicians managed to keep three people alive and healthy for five months, with all oxygen as well as food derived from hydroponically grown crops such as wheat, carrots, beets, and "sedge nuts."[24] Phones and porthole windows allowed the Bios-nauts their only access to the outside world. While their urine was reintroduced into the wheat crop, dried feces was stored, and other biomass was removed to prevent increased $CO_2$. The Soviets also used a thermocatalytic filter to remove "organic admixtures" in the air.

John Allen, Ed Bass, and the Synergians sought a less sterile environment—in fact they wanted to create a miniature paradise—if one undergirded with a "technosphere" to maintain life systems. They were inspired, in part, by the studies of University of Hawai'i professor Clair Folsome who sealed samples of seawater in one and two liter bottles and was able to observe new colonies forming as different waves of microbes established regimes in these tiny ecosystems—some of these sealed sample jars maintained a viable ecosystem decades later. An abstract from one of Folsome's articles from 1988 noted, "Materially closed microbial ecosystems represent model life support systems for the future human habitation of space. These ecosystems when subjected to a constant energy flux seem to be reliable and self-sufficient systems for recycling of biologically produced carbon compounds."[25]

The Synergians also paid homage to the "Gaia Hypotheses" of James Lovelock and Lynn Margulis, which can be reduced, in its most sensational form, to the notion that the entire Earth acts as a living organism with a continually evolving symbiosis between organisms and environment. Like Vernadsky decades earlier, Lovelock and Margulis traced how animals, plants, and microbes create a homeostatic relationship in a changing environment—requiring "coevolution." New Agers were particularly taken with the Gaia hypothesis. Margulis's thumbnail explanation of the Gaia hypothesis, "Life transforms to meet the contingencies of its changing environment and in doing so changes that environment. By degrees the environment becomes absorbed into the processes of life, becomes less a static, inanimate backdrop and more and more like a house, nest, or shell—that is, an involved, constructed part of an organic being."[26]

From Folsome, Vernadsky, and Lovelock and Margulis, the Synergians developed the guiding vision that biosystems are dynamic yet self-regulating and perpetuating. As Allen put it, "I wanted to make the Biosphere 2 experiment prove or disprove once and for all my prediction that a biosphere was a highly adaptable, basically self-organizing life system that could prosper under conditions widely different from actual conditions on Earth."[27] Throw enough species together, in fact, pack the biosphere with species, and unlike God or the Devil, you did not have to know all the details, you could just let the system self-organize. You could create a far-more complex version of Folsome's two-liter flasks—and, preferably, one amenable to human life.

At the same time that Allen was proposing to enclose an entire ecosystem, including billions of unknown microbes so that it could self-organize, scientists in Europe were taking a contrasting approach to creating an autonomous ecosystem. They sought to isolate the minimum number of species (mainly microbes) necessary to maintain basic food crops and sustain oxygen and water supplies; they approached the problem of human survival in space not with a large biome as a model, but instead that of the biochemical laboratory. This project, still ongoing in 2019, MELiSSA (or Micro-Ecological Life Support System Alternative) looked like a series of tube-interlinked steel tanks.

MELiSSA's goal was to turn biological byproducts, whether gases, vegetation, or animal and human waste, 100 percent back to usable resources such as clean air, water, and food. And, as it turned out, the most difficult ecosystem product to work with was human feces. Although it is apparently flung overboard from the ISS, an isolated autonomous community cannot afford to leak resources and maintain itself through the centuries. The highly clinical and compact MELiSSA model has the advantage of a system that could actually fit on spaceships or space habitats.[28] Biospherians were trying to imagine a semi-paradise where people might want to live. MELLiSA's scientists and the Biospherians had quite different definitions of "ecotechnics."

Reflecting a romantic approach, Biosphere 2 would be decidedly more grandiose than the Soviets' 315 cubic meter Bios-3 facility or MELLiSA. More in the mode of Kubla Khan decreeing a stately pleasure dome, aesthetics and poetry vied in importance with science. On a concrete and stainless steel base, the Synergians sealed off a spectacular greenhouse, combining elements of a stepped pyramid, the barrel vaulting of old European train stations, and Fuller domes. These glass structures encased an imported coral reef, a half-acre rain forest with a waterfall, a desert biome, a savannah, a marsh, a high desert "fog forest," and a half-acre farm, as well as the insects, microbes (they did not bother cataloging, just brought the soil, water, and muck), and birds indigenous to these biomes. They also introduced farm animals such as chickens, pigs, and goats—and, as it turned out, hordes of uninvited ants and cockroaches indigenous to the Arizona desert.

Allen and the Institute of Ecotechnics long had cultivated key scientists and intellectuals at yearly conferences. Participants included ecologists Eugene Odum and his brother Howard Odum, geophysicist Keith Runcorn, ethnobotanist Richard Schultes of Harvard, biologist Clare Folsome, astronaut Rusty Schweickart, philosopher Richard Dawkins, biochemist Lynn Margulis, and artists and writers such as Ornette Coleman and William S. Burroughs. Hands on design of distinct biomes for Biosphere 2 went to Tony Burgess of the U.S. Geological Survey (desert), Ghillean Prance of the New York Botanical Gardens (rainforest), and Walter Adey from the Smithsonian (coral reef).

At the outset, Biosphere 2 had scientific credibility. Carl Hodges, an environmental scientist at the University of Arizona deemed the project "totally doable."[29] Others, though excited by the project, were cautious. Howard Odum commented, "on the basis of previous work, I have to say there's a great possibility you will wind up with a mass of green slime, in a year."[30] The Odums, who had led large experiments on the after-effects of nuclear tests on Eniwetok Atoll in the Pacific, appreciated the grand "integrative" systems approach of the Biospherians. Adey, concerned for his reputation, left the project before the Biospherians entered in September 1991. Others, like Burgess, stayed on.

Initial press mixed enthusiasm with bemusement. The effort was called "bold" and "audacious." The human and mythic dimensions were played up over the scientific, and many speculated on the potential sex lives of the eight Biospherians. The Biospherians described how they would enter a new world, or leave Planet Earth by stepping into the facility for two years. Allen made a rather bizarre comparison between Biosphere 2 and the cyclotron—insisting that what the cyclotron was to particle physics, Biosphere 2 would be to, well, ecotechnics. Ecotechnics Institute director Mark Nelson argued that as voyages to outer space had offered astronauts unexpected spiritual experiences related to the "Overview Effect" and a new awareness of human connection to and responsibility for planet Earth, Biosphere 2 would provide a similar spiritual experience via the "innerview effect."[31]

The decision to include under its stately glass lid biomes such as a desert or savannah or even a coral reef was more poetic than scientific. John Allen, and Biosphere 2's other designers, such as Margret Augustine ("Firefly"), were creating their vision of an earthly paradise, an enclosure that could keep people happy for years. *New York Times* reporter Bruce Weber referred to it as "Noah's Ark: The Sequel." *Chicago Sun Times* columnist Bob Greene, in "American Dream Goes Under Glass," discussed the Reagan era trend of "cocooning" or retreating into a private world of comforts, and suggested this trend was epitomized in Biosphere 2. Many of Greene's peers expressed envy for the Biospherians who could leave the rat race for a two-year retreat in mini-paradise. Desert expert Tony Burgess commented, "The metaphor of the Garden of Eden was never explicit, but nonetheless inescapable . . . We would return to the Garden, return to grace with Gaia, through biosphere gates."[32]

Enthusiasm for Biosphere 2 dampened when *Village Voice* reporter Marc Cooper, the spring before the Biospherians entered, wrote an exposé insisting the project was the work of a dangerous and bizarre cult with zero scientific credibility. The gist of the exposé was apparent from its New Journalism title: "Take This Terrarium and Shove It: The Media Loves It, Yale Loves It, Phil Donahue Thinks It's Neat, the Smithsonian Lends Its Name, Scientists Take Its Money—So What If the Biosphere 2 Is Really Run by a Wacko Cult? Don't You Want to Go to Mars?" Cooper interviewed former members of Synergia, consulting scientists, and the anthropologist Laurence Veysey who had studied the Synergia Ranch community during a five week visit in 1971.

Veysey mixed admiration with dread in his account of Synergia. Compared to other communes of the era it was highly organized, directed to achieve specific goals, and impressive. Even he found it seductive. Its leader, whom he dubbed "Ezra" managed to inspire an image of the group as a vanguard that could change history. Yet it functioned as a tyranny and was only semi-benign. Veysey reported that Allen's "domination of the group is open and for the most part undisguised." Commune members sought his approval. He might either support or verbally rip into an individual with an ego-destroying fury that was "awesome and chilling."[33] Mistakes on projects were treated as direct sabotage of the commune effort; when a culprit was exposed, Ezra tunneled into their vulnerabilities, and often threatened them with expulsion. Synergia Ranch resembled more closely a Puritan village than a hippie Eden. Veysey became "keenly aware of the authoritarian dimension to this community's way of life."[34]

According to the *Village Voice* article, Allen was not only verbally but physically abusive. He quoted one former member who said, "When he hits you it's a compliment." Allegedly, Allen frequently verbally abused and even punched and kicked billionaire Ed Bass as he developed his "hypnotic hold" over him. When Cooper questioned Allen, the Biospherians' leader mainly denied the charges, but said that Bass's problem was that everyone deferred to the billionaire. Allen was providing a psychological corrective and claimed he never dished out even 25 percent of the kind of abuse he had experienced as a recruit in military boot camp.

The comparison to boot camp is apt, as brutal initiations, breaking down a person's ego, do suggest indoctrination to a cult-like group. Cooper pushed the premise that Allen's "eerie doomsday dogma makes him much more the Jim Jones than the Johnny Appleseed of the ecology movement."[35] Cooper went on to detail Synergia rituals, mainly culled from Veysey, to prove they were seriously "wacko," such as meditation sessions, group chanting, silent meals, psychodrama-esque encounters, and primal howling.

Cooper also prodded scientists to trash talk Biosphere 2. In their search for expert help, the Biospherians had donated nearly $5 million to the University of Arizona's Environmental Research Laboratory. One of these scientists commented that the original plans for Biosphere 2 were "laughable, idiotic designs" with "no value of any kind." Based on Veysey's account, Cooper concluded that the project "got off the ground as part of an eco-cult intent on surviving the next holocaust."[36] Cooper also lamented that what the Biospherians were doing was not "science but science fiction." This last critique could be applied to much of the space research community, which has some crossover not only with science fiction—but also with myths of redemption and dreams of immortality. Nevertheless, after reading Cooper's article it was tough to accept Allen's romantic account of his history and that of Synergia Ranch.

The controversy swirling around the project only heightened interest in the launch or "closure" date of September 26, 1991. In the *New York Times*, it was noted that despite the organizers' effort to draw a parallel between Biosphere 2 and a NASA project, "The researchers themselves displayed, not the low-key, can-do attitude of astronauts, but the fervor of ecological cultists."[37] Yet, as if from another dimension, another *Times* reporter, Carlyle Douglas, treated it as a light-hearted human interest story, "It's a zoo! It's a greenhouse! No, it's Biosphere 2." Douglas went on to quote one Biospherian's answer to the "big question" on everyone's mind: who would have sex with whom? Biospherian captain Sally Silverstone ("Sierra") said, "We've got four men and four women, so I don't think you can discount the possibility of sexual encounters."[38] Roy Walford, the oldest Biospherian at age sixty-seven, the group's physician, and a gerontologist at UCLA, commented to reporters, "This is not a monastery."

The day before closure, a sweat lodge was built on the hillside, where the Biospherians cleansed themselves. In the evening a dance party was thrown, two thousand strong, fireworks exploded, and Timothy Leary, perhaps extolling life-extension theories, mingled with other celebrities and well-wishers on the lawns.[39] The next morning started with a Buddhist sunrise prayer; the Biospherians in uniforms and makeup were then filmed for morning television talk shows. During a final ceremony, Bass and the Biospherians gave speeches.

After referring to Galileo, Leonardo da Vinci, and the Wright Brothers, Bass stated, "Today we gather at a new instrument . . . to explore the mysteries of our very own home planet . . . Biospherians, good friends, bon voyage, and fly your spaceship well, that all of humankind might fly its spaceship Earth better in the future."[40] Before entering the sealed world, Mark Nelson said, "I have had a vision that man could change from being a desert maker into a builder of oases."[41] Lakota elder Dan Old Elk and a Huichol Indian woman then escorted the eight adepts to the airlocks, and to great cheers, they vanished from the larger biosphere.

Problems that threatened survival soon ensued. Three months after closure, sensors indicated that oxygen levels had dropped 15 percent. The Biospherians were feeling the effects. Biospherian Jane Poynter commented that the system was "self-organizing us right out of the picture."[42] For their safety, 600,000 cubic feet of fresh air, 10 percent of the enclosed atmosphere, was injected in early December. Critics assailed the project's integrity after this and a later disclosure that the Biospherians had quietly, just prior to closure, installed $CO_2$ scrubbers like those on submarines to keep the air breathable, especially during the winter months when photosynthesis activity would slow. Even the news that Biosphere 2 came pre-stocked with a two-month supply of food was derided. Eventually it was determined that the Biospherians' brag that it was the most air-tight structure ever devised proved true—apparently $CO_2$ had been absorbed into the fresh concrete of the base walls, spars, and artificial cliffs, which had not entirely cured when the structure was sealed. (This trapped $CO_2$ could not be recycled via photosynthesis to oxygen.)

The press shifted to hostile. Joel Achenbach's "Bogus New World," came out in the *Washington Post* after the $CO_2$ scrubbers were revealed. Achenbach profiled a computer software expert who had quit the project, saying, "Basically what you have now is a submarine with plants in it." Lynn Margulis, once a fan, noted, "It's not science at all . . . It's a venture capitalist project which is intrinsically fascinating, but it's not scientific in any way whatsoever." Walter Adey, of the Smithsonian, explained he had resigned because Bass and the designers had taken "a Disney approach." (He was particularly disgusted when a viewing window for tourists was ordered for the tropical reef.) The *Post's* Achenbach asked the project director Margret Augustine, "Is this whole thing some kind of ecoscam? A biofraud?" Her response was that they were not taking public funds and therefore could not be defrauding anyone and added, "Achievements like this don't just happen."[43]

Achenbach also talked to the project's remaining defenders. Thomas Lovejoy, an officer at the Smithsonian, recommended more scientific oversight. Stephen O'Brien, a geneticist with the National Cancer Institute said, "There are traditional ways of doing science and there are nontraditional ways. . . . What I see in the Biospherians and all the people working there is a tremendous enthusiasm for exploring the unknown." Achenbach concluded, "This, then, is the paradox of Biosphere 2: It is both awful and wonderful. And if Walter Adey is right, it can be saved." Following Lovejoy's advice, Bass brought in a Scientific Advisory Committee. The move created tension as the Biospherians debated allegiance to Allen, the project's official scientific director, or to the Scientific Advisory Committee.

As for the ecosystem's health—the bees, butterflies, and hummingbirds did not last, nor did nineteen of the twenty-five vertebrate animals, including many of the fish. Cockroaches and ants overran the kitchen (as did feral pigs, which the Biospherians promptly slaughtered). The ocean reef was having problems with acidification from $CO_2$. However most of the biomes and 3,800 species of plants and animals survived, including many of the "bush babies" (galago monkeys) and land tortoises. By the end of the two-year project, the rainforest maintained about 60 percent of its original species. (Twenty years later, as in one of Folsome's seawater flasks, tropical fish still survived in the waters of the reef although it had not been maintained for decades.)

As for the Biospherians, UCLA gerontologist and former Situationist Roy Walford, a crew member, carefully monitored his peers' weight—they were losing a lot—but he believed their diet was safe; they were undergoing "healthy starvation." In cosmist mode, Walford was interested in prolonging human life, and had authored *The 120 Year Diet*. His research with laboratory animals led him to believe a low carbohydrate diet might well add to longevity.

All was not perfect health-wise. Walford noted with concern that levels of toxins such as PCBs and DDE, a decayed form of DDT, were high in their blood, apparently because as they "starved" such poisons were released from their body fat. By the second year the Biospherians had become somewhat better farmers and subsisted on foods such as sweet potatoes, beans, beets, peanuts, rice, and bananas, but food deprivation remained a central issue. As Jane Poynter wrote, when you are growing your own wheat, and making your own cheese "it takes four months to make a pizza." Linda Leigh, head botanist, had friends fax her menus from outside. Colleague Sally Silverstone handled her food deprivation by writing a cookbook, *Eating In: From the Field to the Kitchen in Biosphere 2*. Recipes included, "Beach Blanket Bean Burgers," "Banana Bean Stew," "Biospherian Rice Pudding," and "Biospherian Baked Doughnuts"—oil was too hard to come by to be used in Biospherian cooking.

Psychologically, all was not well. Although Roy Walford continued to mount art happenings under the dome, and arranged, via telephone hook-up, jam sessions between Biospherians and musicians at the Electronic Café in Santa Monica, the rituals and bonhomie that bonded the Synergians outside the bubble had faded with the oxygen. Jane Poynter titled one chapter of her memoir, "Starving, Suffocating, and Going Quite Mad." The eight Biospherians formed two opposed camps, one loyal to "Mission Control"—that is to John Allen and

Margret Augustine. The other group wished to adhere to the advice of the Scientific Advisory Committee.

Poynter noted, of her faction, loyal to the science committee, "The four of Us all felt we had become the underclass in our Dystopia, left out of any major decisions, and at times stripped of responsibility."[44] It got ugly. Team member Linda Leigh wrote in her journal, after two months, "This is the only situation I have even been in that would drive me to drink, except there is no drink here."[45]

Such strains in groups in isolation—whether submarine crews or scientific teams in Antarctica—or astronauts on space stations are apparently common. "Heightened irritability" is one baseline syndrome. A Navy study from the 1960s indicated that as a stay at an Antarctic research station continued, rates of insomnia, depression, anxiety, hostility ("feeling easily annoyed or irritated," "feeling critical of others") all increased.[46] During a mission to Antarctica in 1979 that included sixteen men and one woman, one man, learning of his father's death, went berserk, smashed dishes, chased a rival for the lone woman's affections with a wooden board in hand, and then ran out into the minus 71 degree Fahrenheit landscape—to return later with frostbite. A journalist at the station noted, "Seemingly minor personality quirks or even a slight change in the weather are magnified out of all proportion and can have unpredictable effects on the group's psychological well-being, and conflict flares easily."[47] This took place at a science outpost with plenty of oxygen, equipped with a gym, Jacuzzi, full bar, and wonderful meals. A NASA specialist at the eventual Biosphere 2 debriefing, said, "You guys are textbook."[48]

Strife continued. The Biospherians developed a research plan for the advisory committee to review but when it was submitted all the "science [committee] loyalists," including Walford, the group's only PhD, had been written out of the proposal. The two groups stopped speaking to or even looking at each other. Poynter reported that members of the other group, on one occasion, even spat at her. To the science loyalists ("Us"), those faithful to Mission Control ("Them") seemed insane. A Biospherian such as Jane Poynter, AKA Harlequin, once a dedicated protégée of Allen now found that she hated and despised him and Augustine. She needed psychological counseling, suffered from serious depression for the first time in her life, and experienced hallucinations.

The Scientific Advisory Committee offered praise for the "vision and courage" behind the project, but questioned the veracity of data being obtained, and strongly recommended that Allen be replaced as director of research. During this power struggle, conflict between the two factions heightened under the dome. Margret Augustine at Mission Control wanted Poynter thrown out of the project after she confessed to the project's public relations head, Chris Helms, that they had been eating seed stock such as beans to keep from starving. This led to screaming matches between Augustine and Helms, with John Allen, apparently, curling up in the fetal position in the office. The stress, clearly, was not limited to those under glass. Augustine scolded Poynter. Allen telephoned the heretic Biospherian, sobbing about his failings. The entire team managed to stay in a room together and came to the decision that Poynter would remain. They were fourteen months in.

When the Biospherians emerged, after two years, many of them were taking acetazolamide, an altitude sickness alleviant, to help them cope with low oxygen and $CO_2$ levels four times those found in Earth's atmosphere. The Biospherians had tired of all the pomp and cameras. Jane Poynter remembered waiting through biologist Jane Goodall's long speech and muttered, "Jane, let us apes out of the cage!" Then they were marched out, in blue uniforms that no longer fit them that well—each had lost an average of twenty-five pounds—to the sound of a symphony orchestra, applause from an audience of about two thousand, as well as "much New Age oratory." Walford announced they had demonstrated how to live "closer to the idea of a natural paradise such as the Earth should be and could be." Mark Nelson proclaimed that they had helped uncover "an operating manual for the world."[49] Botanist Linda Leigh—apparently not referring to the restaurant menus friends had sent—claimed, "I have glimpsed paradise."

They had lived on recycled water, grown 80 percent of their own food, required an infusion of 10 percent of their original air, and had survived. If their accomplishments, beyond survival, were hard to sort out, Tony Burgess, the desert biome designer, noted in a 2018 article that, among its accomplishments, surveys of the Biosphere 2 soils led to the discovery of "two species new to science, a protozoan and a soil nematode, indicating that novel communities were forming rapidly from diverse inoculations."[50]

Like one of the adventures in *Don Quixote*, the first "sally" had ended. 400,000 visitors had filtered through the surrounding facility during the two-year run, making it second only to the Grand Canyon as an Arizona tourist destination. The gift shop was stocked with "Bio-Water," Biosphere T-shirts, and "Hug-A-Planet" fabric globes. Verdicts began to roll in. One critic labeled it a "jarring mixture of serious science and New Age eco-kitsch." More damning, an editorial claimed, "If the Biosphere 2 experiment were a horse, it would have been in an Alpo can months ago."

The scientific community offered mixed but mainly negative reviews. Biologist Harold Morwitz stated that in the future people would look back "with awe," and a former NASA scientist, John Corliss, deemed it a success as a "shakedown mission."[51] However, William Knott, a NASA scientist questioned the validity of the data from the project, including its reported low "leak" rate, noting "The Biosphere is not what you can call controlled scientific research."[52] Yet another NASA scientist, Robert Macelroy, was dismissive, "They set out to put people in the box for two years, and they did it . . . We don't know what they've done, what their intentions are, whether they've failed or succeeded. It comes down to the question of what were they trying to do?"[53]

• • •

The second Biosphere 2 mission, begun in spring, 1994, turned more bizarre. Prior to the conclusion of mission one, investment banker

Steve Bannon—who later gained notoriety for his advisory role to Donald Trump and his reign as chairman at *Breitbart News*—was called in to oversee the project's finances and trim expenses. In the middle of mission two, with encouragement from Bannon, Bass's lawyers secured restraining orders against John Allen, Margret Augustine, Mark Nelson, and Marie Harding, Allen's former wife. At the time Allen was in Japan and Augustine in Canada.

On April 1, 1994, armed sheriffs descended on Biosphere 2 headquarters, and installed new locks on all the doors. Off-duty police officers patrolled the grounds. With Bass's blessing, Bannon declared himself CEO. The new crew of Biospherians under the dome, when told of the change in management, thought it was an April Fool's Day prank. When Chris Helms, head of publicity, was informed, he sobbed with relief—obviously traumatized after months working under Allen and Augustine.

More lunacy ensued three days later when two Allen loyalists, and members of the first Biosphere crew, Abigail Alling ("Gaie") and Mark van Thillo ("Laser") broke safety windows to avoid air pressure problems and opened the airlocks on Biosphere 2. They fled, chased by helicopters, and telephoned the Biospherians still under the dome, declaring them liberated. None left. Gaie reports that she was devastated to see bankers take control, fearing they would endanger lives and the project. She commented that, at the time, she "still felt more attached to the Biosphere 2 than Earth."[54] Her mother declared, "She's a victim of mind control. The people who have all been kicked out are members of a cult."[55] The duo was eventually arrested then freed.

All these story lines, suitable, as it turned out, for the bad Pauly Shore movie, *Bio-dome* (1996) and the T. C. Boyle novel *The Terranauts* (2016), made the project hard to take seriously. Yet the facility interested institutions such as Columbia University and the University of Arizona, planning research on climate and ecology. To reclaim legitimacy, officials from these universities hyped the projects their own researchers were conducting and discounted the reign of the Synergians.

But as of this writing, in 2019, with robots crawling across and orbiting Mars, Japan's Hayabusa2 spacecraft retrieving material from a near-Earth asteroid, and Jeff Bezos announcing "It's time to go back to the Moon—this time to stay," attention once again is on long-term space settlement. In this context, Biosphere 2, as originally conceived, no longer seems so wacko. Although NASA and other space agencies have developed high-tech artificial ecosystems, controlled life support systems like MELiSSA are meant for space missions, not for big populations or long-term living.

Biosphere 2, in Allen's view, was designed as a one-hundred-year project. And its image has improved. A 2018 article in *Business Insider* argued that the Biosphere 2 experiments of 1991–94 offer an excellent glimpse into the difficulties of creating a long-term, sustainable artificial environment on Mars or elsewhere in space. Psychologically balanced people are not going to want to live for long in steel containers like those of the five-day Lockheed Lunar Base experiment. Jeff Bezos's 2019 mock-ups of O'Neill colonies were stocked with seemingly natural landscapes—including a feeling of wilderness.

For decades after the two missions, the new Biosphere 2 management tried to disassociate the project from its original mission—during the Columbia University days, for example, one participant likened this legacy to having "dog feces on one's shoe." Now that crud, on closer inspection, might have a kinship with Martian regolith. John Adams, deputy director of Biosphere 2 commented in 2018 that we were "nowhere near ready" to construct colonies in space, but added, "the lessons learned from the original biosphere and the mission are just tremendous, and we could make a much better go at it a second time around."[56] The spaceship may launch again.

CHAPTER SEVEN.

# MAKING SPACE FUN AGAIN

• • •

Percent of U.S. adults who say they personally would be interested in orbiting the Earth in a spacecraft. Not interested: 58 percent. Interested: 42 percent.
—Pew Research Center, March–April 2018

"It's been called the 'James Bond' of space suits by some outlets, following earlier comments from Musk that he wanted it to look 'badass.'"
—*Forbes*, August 27, 2017[1]

If John Spencer has his way, a three-hundred-foot dome of reddish rock will loom above the Las Vegas strip by 2022. The project, which Spencer has been nurturing for a decade, is Mars World, a $1.9 billion space-themed amusement park. With projections on the enclosure to mimic Martian skies, Mars World will include a luxury hotel, a casino, simulated low-gravity entertainment, and opportunities for visitors to don spacesuits, take rover rides through simulated Martian desert, or wander into Martian nightclubs built into the rock of the "impact crater rim" that surrounds the colony.

Spencer is a space architect, the founder of the Space Tourism Society and, ever since seeing Stanley Kubrick's *2001* as a child, a true believer in humanity's space future. He is a dreamer, but as he likes to point out, $250 million in his designs have been implemented either as space projects or Earth-based entertainment. He received two awards from NASA for his design work on the International Space Station (back when it was called Space Station Freedom), planned the Space World pavilion interiors at the South Korean Science Center, and has worked as a consultant for the entertainment industry. ("You know the interior of the spaceship that took the hero to Pandora in *Avatar*? That was from a design I developed with Buzz Aldrin for another James Cameron movie, on Mars, that hasn't been produced yet.") His company has several mottos for Mars World: including "Like No Place on Earth," and "For the Explorer in All of Us."

In a sense, Spencer has always lived in Mars World, that is, in the zone that blends genuine space design with themed entertainment. He insisted that astronautics and what he called "simnautics" overlap. "There has always been a connection between entertainment and real exploration. Exploring is fundamental to us. Exploring gives us a positive view of the future—or why bother? And it can generate amazing stories. Space happens to be one of the stories of our time. Hundreds of years ago, it was crossing the oceans. Or reaching the North or South Pole. Now it is space. Explorers want to tell their story and the media wants to document wild stuff. There is a synergy. And these stories are good for kids. It's exciting. It gives us a sense of nobility. It's important."[2]

He insisted we will go to Mars, for the story, if nothing else. He noted that "whenever we think of Mars, it is near term, with four or five astronauts and they're struggling to survive and they're all dirty and not having fun." He wanted to shift that image. To an audience in 2016, he suggested, "even the tragedies that will happen will keep Mars in the public awareness. . . . the Martian theme will psyche a lot of people who want to experience space."

While he also has devised a Mars camp for children, Spencer visualized Mars World as skewed toward adult entertainment, since 95 percent of tourists to Las Vegas are twenty-one or older.[3] He projected for Mars World forty thousand such visitors a day (about one in ten of Las Vegas's total daily visitors). Mars World will not be hot and sweaty. Its immersive setting will conjure a colony inside an impact crater near Olympus Mons in the year 2088. "Settlers," he explained, "came to the area and slowly expanded around the rim. On actual Mars you'd need to be underground—because of the radiation. There would be no dome. The surface of Mars is not a healthy place and a dome would not be a good idea. To build it would take time, energy, materials. It could be sabotaged. And why try to maintain such a large artificial atmosphere? Tunneling into a crater rim makes more sense."

But, as he told me, "The core part of Mars World is not just the building and the technology, but the invention of a Martian culture of the future. Imagine," he said, "Burning Man on Mars . . . Our Martians, in our thinking, are pretty out there and this fits Las Vegas, as it is a pretty wild place."[4] Mars World will include "tons" of robots, as well as butterflies—actual and robotic. ("Martians love butterflies," he insisted.) In the crater walls will be airlocks, caves, rovers to traverse Martian landscapes, taverns serving Martian drinks, and actors sporting futuro-Romanesque fashions. As scholars would say, Spencer was striving to create a "liminal space," to open people to a different awareness.

Key to the design was the "space-time vortex" visitors pass through. He compared it to entering Disneyland which has what he called a mild transition. "In Mars World the transition will be more intense. You walk up a ramp that is curved so you can't see where it leads, with lots of light displays. After a slight incline you make a turn. Then it feels different. You have left Earthport for Marsport." The closer you get to the center of the crater dome—on three glass-enclosed tunneled walkways—the "more intense" the experience becomes—presumably with Cirque du Soleil style displays of near zero-gravity acrobatics and sports.

Sightlines will prevent guests from seeing structures beyond the sloping crater wall—which will rise one hundred feet or more. The nighttime skies, true to Mars, with its slight atmosphere and great distance from the Sun will be ablaze with, Spencer said, "a bazillion stars. This we'll be accurate with. With computer systems we can mimic how the skies will change as Mars moves through its orbit during its long year. Most visitors won't care but we will cater to the one percent who do care." He also planned the dome to be as "self-sufficient, energy efficient, and energy productive as possible" to model aspects of an actual Mars colony. He suggested that overnight guests at the hotel who wake up will really feel "there." While he wanted the night skies to be accurate, he was not going to switch the clock to Mars's 24-hour-and-37-minute-long day. Nor would he be able to mimic the gravitational pull on Mars which is about one-third of that on Earth. "But it can be simulated in various ways."

When I ask how he got hooked on space, he answered, "In the late 1970s, when I was studying architecture, I was introduced to Gerard O'Neill's space colonies proposal. That was a moment where I knew what I wanted to do for the rest of my life. I could do what I love as an architect and designer and work on space. There really are not too many of us. There are about a half dozen space architects in the world today." He credited the O'Neill-inspired L-5 societies with jump-starting the private space movement. "The idea that private citizens could band together and have an effect on space just hadn't been considered before." As an apprentice architect, he also worked with the firm that designed the space frame for Biosphere 2, which he called "a much more intriguing project than it has since been made out to be."

Spencer was not a man with one fixed idea. Mars World would be for the masses, but for the wealthy, he has designed a luxury space yacht, *Destiny*. The "profit and loss" business model can be set aside, he argued, when creating projects for the ultra-wealthy. Earth's luxury yachts, besides being "amazing pieces of technology," exist to confirm social standing, bolster ego, and provide branding and mystique. "You can only add so much art, so many submarines or helicopters or chefs to your yacht before you are outdone. The answer for someone like this is to be the first to create an orbital yacht. My spaceship *Destiny* will be three hundred feet long, have room for ten passengers, six crew members, and many, many robots."

He envisioned space yachts, which could cost five billion dollars to build and half a billion dollars a year to maintain, as becoming a reality within twenty years. Purchasers might be an individual or a corporation. In his universe, space yachts, yacht races, and other space sports will eventually be joined with space ocean liners, as well as a fleet of Space Guard support ships, resorts, and settlements. "We will have to work out how to prepare and clean up five-star meals in space." He paused, "You know, the ultra-wealthy really belong to an alien culture."

Spencer's pun underscored a paradox. There has been surprisingly little crossover between advocates of spacefaring and aficionados of UFOs. The likely distinction is power—with one exception in former President Jimmy Carter—few in power openly express interest in UFOs; generally, such tales percolate from the bottom up and, on a mythic level, ally the powerless with cosmic forces. One marked exception is Robert T. Bigelow, Las Vegas businessman, founder of Budget Suites of America and Bigelow Aerospace, and a prime mover in the budding space tourism industry.

Bigelow, a close double for the sharp-eyed western movie actor Lee Van Cleef, has no doubt that aliens have visited Earth. He grew up hearing his grandparents' account of seeing a UFO approach their car then take off at a right angle when they were driving on a canyon road to the north of Las Vegas. His curiosity was piqued: at age twelve he resolved to earn the fortune that would allow him to enter the space business. Bigelow, who has since had his own UFO encounters, believes in a continuing alien presence on Earth, reflected in the logo on the side of the Bigelow Aerospace facility in Las Vegas, a calligraphic version of a big-eyed "grey."

With the help of Nevada Senator Harry Reid, from 2007 to 2012 Bigelow gained $22 million in Pentagon funding for a smaller version of the Air Force's Project Blue Book of the 1950s and 1960s. He called in experts to review video and audiotapes and to study alleged UFOs remnants brought to Bigelow hangars.[5] When *Sixty Minutes* reporter Lara Logan asked if his belief in extraterrestrials undermined his credibility, Bigelow responded, "I don't give a damn. It doesn't change the reality of what I know."[6] In another interview, cited by the *Times*, Bigelow noted that unlike other countries that openly investigate UFOs, the United States was "being held back by a juvenile taboo."

In 2003 Bigelow, who is in the hotel business, licensed and improved sidelined NASA technology for an inflatable space habitat. His company launched two small inflatable "Genesis" space-habs in 2006 and 2007. In 2016, a SpaceX rocket delivered the Bigelow Expandable Activity Module (BEAM) to the ISS, where it was attached and successfully inflated (although it took two tries on two days rather than the few hours expected). The Beam provided 16 cubic meters (565 cubic feet) of space; Bigelow plans to launch a larger fully autonomous space station, its B330, providing 330 cubic meters (11,654 cubic feet) of space into low Earth orbit by 2021. The B330 can support six people, while linking two B330s together will double its capacity. Bigelow will remain in the real estate business, renting these modules out to Earth-based customers, whether as orbiting science laboratories, corporate conference centers, or hotels.

• • •

While science writer James Gleick has dubbed space tourism "cheap thrills for plutocrats"—and how else can you describe a corporate retreat in Earth orbit?—proponents point out that sending the ultra-wealthy to hotels in space is an opening wedge. It was also, in a Chamber of Commerce sense of the phrase, all-American. As one advocate phrased it, "Space tourism is of the very essence of what it is to be American, in that it is designed for 'the pursuit of Happiness,' and involves taking risks, pushing back boundaries, and making a buck in the process."[7] The strongest argument against the "cheap thrills" argument: wealthy space tourists just might experience a spiritual transformation or cognitive shift termed the "Overview Effect."

Author Frank White developed the concept of the Overview Effect after interviewing astronauts who reported experiencing moments of profound insight and transformation when off planet. They described a feeling of deep connection to an Earth without borders, to humankind, and to the universe. Several became deeply religious, while others found their space experience heightened their religious sensibility whether of the New Age or orthodox variety. Apollo 14 astronaut Edgar Mitchell, who later founded the Institute of Noetic Sciences in

1973, described having an "explosion of awareness" and an "overwhelming sense of oneness and connectedness . . . accompanied by an ecstasy . . . an epiphany."[8] Soviet cosmonaut Boris Volynov, expressed an enduring psychological shift, "During a space flight, the psyche of each astronaut is re-shaped; having seen the sun, the stars and our planet, you become more full of life, softer. You begin to look at all living things with greater trepidation and you begin to be more kind and patient with the people around you."[9]

White founded the Overview Institute in 2008, along with astronaut Edgar Mitchell (who died in 2016), *2001*'s special effects creator Douglas Trumbull, New Age philanthropist Barbara Marx Hubbard, and other artists and space entrepreneurs. Its purpose, "From space, the astronauts tell us . . . the conflicts that divide us become less important and the need to create a planetary society with the united will to protect this 'pale blue dot' becomes both obvious and imperative."[10] They insist that space tourism should not be dismissed as mere diversion for the wealthy, but a tool for profound political change.

In the *Right Stuff* Tom Wolfe critiqued astronauts such as John Glenn as so thoroughly trained in simulations that the experience of being in space had no ability to inspire awe—yet if White was correct, spaceflight, rather as advocates of L.S.D. and psychedelic drugs once had promised, could turn engineers and military personnel into poets. The Overview Effect has become a big selling point in luxury space tourism. Paying millions to experience a shift in awareness—perhaps another example of therapeutic culture run amok, was, arguably, an opportunity for redemption for the privileged class. In addition to Bigelow Aerospace, many others were in the field, including Orion Span, which planned to open its Aurora Station luxury hotel in 2022 with twelve-day stays priced at $9.5 million. Its guests would be required to train for ninety days before they blasted off to enjoy the Overview Effect, as well as "zero gravity ping pong," and "top quality space food."[11]

Another start-up, Axiom Space, whose CEO, Michael T. Suffredini, was the program manager for the ISS for a decade, stressed the nobility of space tourism, that is, the opportunity for tourists to experience the Overview Effect and return to Earth as advocates for world unity. Axiom guests would train several weeks for their space stays. In addition to these private contenders, in June 2019, NASA, hobbled by its $3 to $4 billion yearly budget for the ISS, announced it would open the space station to commercial ventures as well as tourism. NASA's first space tourists might be shuttled to the space station in 2020 at $58 million for a thirty-day visit. ISS tourists will also have additional charges of about $35,000 a day for access to a toilet, food, air, data, power, and medical supplies.[12] At a slight discount, Bigelow Space Industries announced in June 2019 that it would broker one to two month trips to the ISS via SpaceX launches, at about $52 million per customer.

In addition to the lure of adventure, and the possibility of spiritual renewal, promoters long have added erotic hijinks to the mix. In a 2009 interview, Spencer suggested, "Sex, drugs and rock and roll is what's really going to build the space environment . . . Planting flags is old hat."[13] But will space yachts or inflatable hotels be the great honeymoon destinations that space hotel builders and *Omni* magazine once envisioned?

Several books about space have suggested that sex may not be that great in an orbital honeymoon suite. In *Packing for Mars*, Mary Roach cited Johns Hopkins's researchers who observed two harbor seals mating in a swimming pool and noted that thanks to Newton's Third Law—each action has an equal and opposite reaction—it was difficult for the buoyant seals to remain together. Likewise, *Wired*'s David Kushner, spending time at Russia's Star City astronaut training facility in 2008, wrote that male impotence in space was a well-known secret. He quoted, in dialect, a Russian press liaison who told him, "There vas top-secret program of this . . . But the man could not perform. Viagra vill not help."[14]

Spencer insisted this was nonsense and referred me to Alexander Layendecker's studies of human reproduction in space. Layendecker, an Air Force rescue pilot who for years was part of Air Force space launch operations, in 2016 completed his densely-titled PhD dissertation, *Sex in Outer Space and the Advent of Astrosexology: A Philosophical Inquiry into the Implications of Human Sexuality and Reproductive Development Factors in Seeding Humanity's Future Throughout the Cosmos and the Argument for an Astrosexological Research Institute.*

Layendecker told me that he was led to the topic by "happenstance. I had been immersed in the space environment in the Air Force and was simultaneously pursuing an M.A. in public health, and then at the Institute for Advanced Study of Human Sexuality was looking for a dissertation topic. I decided that sex and reproduction in space had not received the attention they deserved; if we're serious about discussions of colonization—having babies in microgravity, on Mars or other outposts of the Earth—then more needs to be learned."[15] His general recommendation was that because of the squeamishness of NASA to study sex in space, a private nonprofit organization, ideally his Astrosexological Research Institute, should be founded for this research critical to human settlement of outer space.

From a space tourism perspective, Layendecker's study of the literature yielded both good and bad news. Sex should be possible, even lively, but reproduction, critical for space colonization, could entail severe health consequences. His assessment included glum early accounts, like that in *Wired* magazine, of male impotence: "Reports on the incidence of this in microgravity conditions by male astronauts are mixed—the astronauts of the Mercury and Gemini programs reportedly confided to flight surgeons that they noticed a 'definite lack of activity in that region.'" This was also backed by evidence of lowered testosterone levels in astronauts "exposed to microgravity conditions." Layendecker then brightened the picture, noting, "Astronaut Mike Mullane reported awaking to erections so intense that they were painful during his first mission in space aboard Space Shuttle Discovery on STS-41-D." Mullane, in his memoir, observed, "Flight surgeons have attributed this phenomenon to the fluid shift that occurs during weightlessness. On the Earth, gravity holds more blood in our lower legs. In orbit that blood is equally distributed throughout our bodies. For men the result is a Viagra effect."[16] Dennis Tito, a former JPL scientist and investment tycoon corroborated this narrative. After spending eight days on the ISS as the first space tourist, Tito announced, "In space, there is no need for Viagra."[17]

In another publication, Layendecker along with coauthor, the physician and Canadian "citizen-scientist astronaut candidate" Shawna Pandya noted that there was no available data from women about their levels of sexual arousal while in space. NASA, apparently, does not ask, and female astronauts, so far, have chosen not to tell. Layendecker suspected that because of the engrained misogyny and sexism prevalent among Air Force pilots and at NASA, female astronauts have not been willing to risk their careers discussing with Jimmy Kimmel or CNN reporters whether they feel sexy in space. To bring home this point, Layendecker described how the career of NASA psychiatrist Dr. Yvonne Clearwater was buffeted in the 1980s, when she led NASA Ames's Habitability Research Group.

For a 1985 *Psychology Today* article, Clearwater authored a sidebar titled "Intimacy in Space" explaining why, as a habitat designer, she favored creating privacy, including sound-proofing, around space station sleeping quarters, "The prospect of having women and men working together in close quarters always seems to lead to questions about sexual activity . . . It seems obvious, however, that a group of normal healthy professionals will probably possess normal healthy sexual appetites. . . . If we lock people up for 90-day periods, we must prepare for the possibility of intimate behavior." She added that the job of her team was "not to serve as judges of morality, but to support people in living as comfortably and as normally as possible."[18]

Her calm assessment created a media nightmare for NASA. Newspapers ran articles with titles such as, "NASA Told to Make Room for Sex in Space," and "Sex in Space: Do You Want Your Tax Dollars to Support This?" According to Layendecker, NASA had to hire a full-time liaison to assure Congress that they were not budgeting tax money to run Kinsey-like sex experiments. Journalist Laura Woodmansee wrote, "From the reaction . . . you would have thought she was advocating genocide."[19] At a space conference seven years later, Clearwater, when questioned about the sex lives of astronauts, responded, "We are not doing any research and it's a non-issue."[20]

Even today, as billionaires with their own space companies promote the multiplanetary future, NASA would prefer not to spend taxpayer dollars on sex studies and be accused of abetting erotic or adulterous activity. Sadly, for the Puritans in space crowd, in 2008 and 2009, with very little public fanfare, four private sleeping quarters were added to the ISS, each about the size of a telephone booth, just as Clearwater's team had recommended in 1984. Of these new crew quarters (CQ), lined with acoustic blankets, a NASA publication indicated "crew members could not hear other crew members outside the CQ when the doors were closed."[21]

To return to Layendecker's good news—Tito and Mullane's accounts made clear that sex could certainly occur in microgravity. But Layendecker and his coauthor indicated that there could be hygiene issues. As they delicately put it, "Externally, water-based fluids cohere together in microgravity conditions due to surface tension; sweat, saliva, bodily secretions, and other fluids could build up during intercourse in microgravity. Practically, this can interfere with adequate lubrication and also be messy."[22] Translation: a "used" honeymoon pod might be the last place in the universe you would want to voluntarily enter.

Further, the physics of sex in microgravity suggested that humans, however willing and able, could have the same problems Johns Hopkins researchers observed in harbor seals. "Beyond arousal, coitus itself becomes much more physically difficult in microgravity, due to an imbalance of forces created by partners with differing masses and centers of mass." But, "successful coitus could be wholly achievable with the assistance of tethers and restraint-type devices, such as Bonta's 2suit, designed to stabilize at least one partner during intercourse in microgravity."[23]

Vanna Bonta, who died in 2016, invented the 2suit after she and her husband discovered how difficult it was to kiss in microgravity during a parabolic flight in 2006. Her 2suit allows partners to Velcro to one another before engaging in intimate behavior. Bonta and her husband apparently tested the suit and simulated having sex during a parabolic flight in 2008 filmed for the History Channel (whose programmers, clearly, have branched out from an earlier taste for World War Two combat footage).[24]

Of graver concern was the viability of pregnancies begun in space—and the health effects on mothers and offspring. Layendecker was floored when he was asked to consult with the Dutch startup "SpaceLife Origin," which, in the fall of 2018, proposed to provide wealthy mothers the facilities to give birth in space, with the motto, "If humanity wants to become a multi-planetary species, we also need to learn how to reproduce in space."

SpaceLife Origin had a multistep plan. First, by 2020, they would implement the SpaceLife Origin Ark, which would be a bank of male and female sex cells, "protecting the cells for any catastrophic event on Earth for decades." (This was Plan B in the movie *Interstellar*.) Next, by 2021, SpaceLife planned to launch "Mission Lotus" in which human ovum would be fertilized in orbit and then returned to Earth and if deemed viable, implanted in a mother. Finally, by 2024, the company would operate a birthing service in orbit, or "Mission Cradle," in which pregnant women, near term, in a one- to two-day mission in low-Earth orbit, "accompanied by a trained, world-class medical team," would give birth. As in other forms of luxury space tourism, parents of such space babies would be wealthy paying customers, presumably motivated to be part of history, and to have children who truly earned exotic names like "Moon Unit 2."

Layendecker was not the only one flabbergasted by SpaceLife Origin's proposal. Microgravity and exposure to increased radiation in space were not trivial. He said, "There are real dangers, hazards to successfully produce healthy babies in that environment. And all sorts of liability issues." Layendecker added, "I advocate starting with mice, rats, but companies like SpaceLife Origin were jumping straight to humans."

In general, sexual reproduction in space will be medically fraught. In a worst-case scenario, the high levels of radiation in space could render males temporarily and females permanently sterile. Russian experiments with rats in satellites revealed that although coupling and fertilization took place, no offspring were born. Once back on Earth, male space rats tended to induce "significant abnormalities" in offspring when mated with partners that had not flown. Layendecker pointed out that there also were concerns about "significant alterations" to sperm and ovum development in space, "raising serious questions about humanity's ability to successfully reproduce off-world."

NASA's cautious planners took note. "A number of NASA researchers have suggested offering cryostorage of gametes for astronauts of both genders, in order to safeguard their ability to reproduce if desired."[25] Layendecker also suggested further study was needed of the likelihood, in microgravity, of ectopic pregnancy, in which a fertilized zygote implants in the fallopian tubes rather than the uterus, as well as endometriosis, which involves abnormal growth of the uterine lining. Both conditions can be life threatening. SpaceLife Origin, it would seem, despite its utopian and profit-seeking goal, was boldly wandering into a minefield for litigation.

In regard to Spacelife Origin and others' ambitions to open the path to off-planet colonization, Layendecker commented, "Frankly there are a lot of people in the space community that we refer to as 'space cadets.' They are incredible and enthusiastic about space. This is fantastic. And commercial ventures are enabling us to get into space. A lot of us care very much about humanity expanding into the cosmos, but we also are realists." Layendecker asserted that until researchers acquire more knowledge about reproduction and other areas of space medicine, the vision of off-Earth settlements needed to be tempered. "There are so many issues of spaceflight science and medicine that we have to solve . . . It's so much easier to do a math problem than to work out all the issues of biology and physiology responding to space."

SpaceLife Origin, to its credit, took the criticism of Layendecker and others to heart. As of the summer of 2019, Kees Mulder, CEO, announced he was suspending its programs. "Serious ethical, safety and medical concerns related to Missions Lotus & Cradle are preventing me personally any longer from accepting any associations with and responsibilities for those two specific missions involving 'embryos, pregnant women and baby's [sic.] in space.'" He added, "In short: 'Better safe than sorry.'"[26]

• • •

A few months after I spoke with John Spencer about his plans for Mars World, in Anaheim, Disney opened its fourteen-acre *Star Wars*-themed Galaxy's Edge. Disney's Imagineers hit on many of Spencer's goals: a tunnel at the entrance leads visitors through the murmur of John Williams themed music and on to the "reveal" of the burnt chalky brutalism of Black Spire Outpost on planet Batuu. The park recreates, for *Star Wars* fans, a "busy, rugged spaceport" full of smugglers, rebels, and locals menaced by Storm Troopers. Members of the rebel opposition proposition visitors to build light sabers in a black market shop, while alien music tinkles from radios, roars come from unseen spaceship launches, and signs abound, lettered in the alien *Star Wars* language of Aurebesh. (An app on visitors' phones provides translations.)

Unlike much of Disneyland, Galaxy's Edge was pitched to an adult crowd, and included audience interactivity and role play ("our guests . . . want more and more to lean into these stories," one designer explained).[27] Staff dispense alcoholic drinks at Oga's Cantina as well as nonalcoholic blue milk, and other "drink orbs" embellished with labels in Aurebesh. One big difference between Galaxy's Edge and the yet-to-be-realized Mars World was that Galaxy's Edge aimed at fast turnover. While Spencer wanted hotel guests to "wake up" on Mars and really feel they were there, Galaxy's Edge guests that summer had four hours before their wrist bands alerted Storm Troopers to remove them from the park.

And while Galaxy's Edge simulated an alien world, as would Mars World, it was a different sort of simulacrum than that which Spencer imagined. It did not promote space exploration, or provide a sense of a real settlement of the future, but rather, a chance to revel in participatory fandom. In the language of semiotics, while Mars World might serve as a signifier for the planet Mars, Galaxy's Edge was the signifier for a fictional world, making it at least two simulations removed. To simplify: Mars exists and Batuu does not. Oddly, the Galaxy's Edge visitor who consumed a "Ronto Wrap" (apparently the meat of a beloved Tatooine pack animal) was indulging in a very real consumer experience many times removed from reality, but more real than that the actors enjoyed while filming the movie. Just as easily, roaming Galaxy's Edge could be filed under the category of expensive play. (The park entry fee is $129; Light sabers cost $200; to build a droid was $100—a relative bargain.)

If we insist that space tourism might lead participants to cultivate a taste for outer space, i.e., serve an educational or inspirational function, then Galaxy's Edge was at the far entertainment end of this "influencing" spectrum. Edging it out as education would be Spencer's Mars World, and the spectrum would continue, perhaps, to NASA offerings for tourists, and then to science-grounded simulations such as the Mars Society's Mars Desert Research Station in Utah—where volunteer crews conduct two-week missions in colder months.

What sort of space simulacrums did NASA offer? At the Kennedy Space Center Visitor Complex at Cape Canaveral (a short drive from Walt Disney World in Orlando, Florida), visitors could sign up for an "Astronaut Training Experience," a five hour, $175 opportunity to practice "docking skills, navigate the unique Mars terrain and experience the sensation of performing a spacewalk in a microgravity environment."[28] The center also offered visitors five-hour tours of "Mars Base 1" for $150. Here a "Rookie Astronaut" could "manage the Base Operations Center on Mars, harvest vegetables in the Botany Lab" and "program robots to optimize solar energy intake." Visitors on a budget for $40 could sign up for a forty-five-minute "Spacewalk Training," or spend an hour in "Mars Exploration Simulator Training."[29] (No charge for the daily opportunity to encounter a veteran astronaut, and take a selfie with someone who has worked in space, likely experienced the Overview Effect, and, because of this, will display infinite patience.)

Imagine now, if you will, a group of rookie astronauts, or simnauts, carefully culled to represent NASA's version of the fittest. Such candidates have to be thirty to fifty-five years old, with a background in science, technology, or the military, and be willing to be locked up for forty-five days in an elaborate simulation. Suddenly we are in the real world of space studies, social science wing, more specifically NASA's Human Exploration Research Analog (HERA) at the Johnson Space Center (JSC) in Houston.

From a pool of about four hundred applicants a year, NASA selects its four crews of four, generally of mixed gender. None can have a body mass index greater than 29, be taller than 74 inches, smoke cigarettes, take medications, have dietary restrictions, or "a history of sleepwalking." Those who qualify then take physical exams and undergo a psychological assessment. (NASA covers travel expenses and pays a per diem higher than that of jury duty—$10 per waking hour, usually sixteen a day.) In 2014 HERA began running seven-day analog missions; in 2015 NASA stepped it up to fourteen-day missions, then thirty-day missions in 2016. In 2017 they began running forty-five-day missions.

In each yearly campaign, NASA has run the same mission with four separate crews to uncover variants in behavior and responses. Beginning in 2015, the HERAnauts took simulated voyages to the near-Earth asteroid Geographos to collect mineral samples. As of May 2019, NASA shifted the simulation to a journey to Phobos, the larger of Mars's two moons.

Successful applicants train for sixteen days before being led inside a Johnson Space Center warehouse. They then enter a two-and-a-half story, 300 cubic foot "spaceship" consisting of two conjoined cylinders: the horizontal "Hygiene Module" connected to the upright main silo "Segment B," daubed with NASA insignias and small American flags.

The habitat, 636 square feet, about half the size of one of Bigelow Aerospace's inflatable modules, although bigger than many an apartment in Tokyo or Manhattan, remains small enough to make its four volunteers irritable. Life in Segment B resembles a theme-park ride set to slow. Along with sound effects and motion inducers to mimic blast off, all the windows have computer screens to simulate a journey into space, revealing starscapes and the eventual approach to the asteroid or moon. As part of the mission scenario, the test subjects take space walks, land on the target, and operate a robotic arm to secure mineral samples.

But this was no theme park ride designed to maximize pleasure. A major point of the missions was to learn how to avoid the near warfare that emerged between the participants of Biosphere 2—when Biospherians split into factions and reached the point that they were spitting at one another. HERA was a "controlled analog," focused on group dynamics and how people react to confinement and isolation.

HERA's goal has been to create teams that can successfully navigate the stresses of long journeys in space, avoid cognitive lapses, and ensure participants do not become defiant with mission control, one another, or otherwise go rogue. (The ideal NASA mission, unlike Joseph Conrad's *Heart of Darkness*, should not end with a lone astronaut crying, "The horror! The horror!") Testing "stressors," factors that make people irritable, was one of NASA's aims.

Many are the physical stressors that inhabitants of spaceships or stations must navigate: *"Limited available space, constant confinement, irregular or unnatural light cycles, changes in pressure, extreme temperatures, unusual environmental hazards (meteorites, radiation, etc.), noise and vibrations, poor ventilation, sterile and monotonous surroundings, physical threat to life in exterior environment, and restricted diet."* But wait, it gets worse! Social stressors—somehow this list reminds me of my years in high school—include: *"The feeling of being over-crowded, the feeling of loneliness and separation from one's normal social group, reduction of privacy, the necessity of forced interaction with a small group of people, dependence on a limited community, disconnection from the natural world, no separation of work and social life, no family life, repetitive and often meaningless tasks."*[30] On an episode of *NASA, We Have a Podcast*, an official boiled down the concern to, "we are getting data to keep crews happy on real missions."

Keeping real astronauts happy has involved potentially making HERAnauts miserable. The asteroid missions involved sleep deprivation, whether acute (keeping up the crew twenty-four hours one day a week) or chronic (limited sleep five nights a week). For the Phobos missions that began in May 2019, the HERAnauts were given less privacy in their crew quarters and "less free volume" in which to roam. To increase stressors during the Phobos journey, the privacy door between the core module and the hygiene unit was removed and a curtain put in place to separate the toilet. Storage cabinets were removed and a false wall added to inhibit movement. Stowage bags were placed throughout to cramp the HERAnauts' mobility. According to Lisa Spence, then Flight Analogs project manager at NASA, many of the stowage bags were filled with shaped foam.[31] (It was at this point in our conversation that I realized I might not be a good candidate for HERA. Spence said there was always a "tug of war" between her team and researchers who wanted to increase stressors.) A maze of these stowage bags in the core created a "switchback," slowing passage.

Spence told me that for the next HERA campaign, beginning in May 2020, "autonomy" would be the key factor studied. Crew members would have more control scheduling tasks, and would be expected to problem solve and handle maintenance, with less guidance from mission control. I asked if this meant there might be mock emergencies, or repairs that crews would have to undertake? Spence answered that, more likely, the focus would be on slower communication between the crew and ground control. "The mission might explore lost communication for periods of time, whether scheduled or unannounced." She added that researchers may devise tools to aid crew members at times of increased autonomy, and when I asked if this might be an artificial intelligence system, a shipboard HAL or Alexa, she was noncommittal and said the tools had not been developed yet. (My question wasn't all that rookie— European astronauts on the ISS now have an AI helper, CIMON 2, complete with smiley face, that floats along with them to chat and offer scientific and emotional support.)

The HERAnauts, then, were lab rats, providing data for social scientists. One such study tracked the 2016 campaign's thirty-day missions to the near-Earth asteroid Geographos. Researchers tested the crews' cognitive abilities on three days to "generate" (or brainstorm as a group about different ways to use a common object), "choose" (analyze a survival situation in an extreme environment), "execute" (perform a motor task), and "negotiate" (discuss an ethical dilemma, for example the problem of a runaway trolley car in which the crew must decide whether to plow into a crowd or shunt onto a side track and kill only one individual). The experiments were run on Day 11, Day 16 (after communications with ground control became slower and crew autonomy increased), and Day 30. Researchers found that the HERAnauts' creative thinking abilities and plans for approaching a survival situation all steadily decreased (perhaps because of sleep deprivation), whereas their ethical instincts remained stable, and their ability to execute tasks as a team—perhaps as they got to know each other's strengths and weaknesses—all improved.[32]

Despite all the stressors negotiated, "yelps" from HERAnauts—all space enthusiasts, some of whom had applied for the astronaut corps before joining the experiment—have tended to be good. "Time flew," one remarked "23 days felt like one week. It went fast." A biology professor turned HERAnaut said, "It's weird not to see the sun and not hear the rain and not feel the wind. But you don't dwell on that because you're so busy doing other things."[33] Richard Addante, on Mission 14 in 2017, noted, "Life is different in a space capsule: There is nowhere to escape to or hide at, and we can't just leave or run away from things you don't like. You are really forced to work together to figure it out, so it made for better teamwork and made us better teammates."[34]

As for the solidity of the simulated reality, Addante commented, "As we had no contact with the outside world, it really gave us the sense of being immersed on a space mission, which we took so seriously that upon egressing, it took us time to re-adjust 'back to earth.'"[35] Spence gave her team props for making it real. "We design a mission scenario, mission tasks, and timetable. The module has the look and feel of the interior of a spaceship." As they near their return splashdown to Earth the HERAnauts hold a media conference. And on exit they are asked how immersed they were in the environment and what brought them out of it. One factor: Texan rain storms. "You hear the thunder and you remember you are in a habitat, in a warehouse, at the Johnson Space Center in Texas and not on a space mission," Spence said.

• • •

While long space excursions—simulated or real—may be a grind, we have to assume that paying $250,000 to experience three to five minutes of weightlessness high above the Earth will be one expensive "cheap thrill." As of this writing, two companies, Virgin Galactic and Blue Origin have been in a race to initiate commercial space suborbital jaunts. In late summer 2019, preparing for flights, Virgin Galactic opened its terminal at Spaceport America, in New Mexico. The furnishings in the terminal relied on a "desert palette" of colors and included "stone-colored" boomerang couches. The overall effect was vaguely reminiscent of spaceport interiors in 1960s science fiction depictions.

When I met Virgin employees manning a table at a space event, I sensed that the corporate culture at the Virgin Group really was as youthful, upbeat, and peppy as Richard Branson sought. Branson's swashbuckling image has inspired the corporate mantra "Work hard, play hard." His twitter bio: "Tie-loathing adventurer, philanthropist & troublemaker, who believes in turning ideas into reality. Otherwise known as Dr. Yes." Branson dropped out of school at age fifteen, and by age twenty had started the mail order record business that became Virgin Records. He has been known for his taste for hot air balloon races, sky-diving, driving a tank through Manhattan, kite-surfing the English Channel, and rappelling down Virgin Galactic's headquarters in New Mexico after the 2019 move from Mojave, California.

Spaceship designer Burt Rutan's Scaled Composites originally designed Virgin Galactic's passenger-bearing SpaceShipTwo (SS2).

Like the rocket planes Air Force pilots such as Chuck Yeager tested from the late 1940s on—a "mother plane"—in this case Virgin Galactic's twin-hulled WhiteKnightTwo—will carry the SS2 to an altitude of about fifteen kilometers. After being released, the rocket plane then will take off to the neighborhood of the Kármán Line, about 100 kilometers above the Earth, where outer space begins. After a brief suborbital flight allowing three to five minutes of microgravity (quick, don the Banta 2suit!), the SS2 will glide back to its landing strip.

Virgin Galactic suffered a major setback in 2014 when during a test flight over the Mojave the air braking system was activated early and a WhiteKnightTwo disintegrated, killing the copilot (the pilot managed to parachute to safety). After five years of redesign and testing, Branson announced that commercial flights would begin in 2020. With any luck the jaunt will live up to Virgin Galactic crew member Beth Moses's description of her "mind-blowing" test flight above the Kármán Line. This former NASA engineer said the flight can give you the sense that "the sands of time of your life have stopped for a moment."[36]

Prior to the 2014 accident, Virgin had sold approximately six hundred seats at $250,000 per passenger. Both these old and newer customers, six per flight, will soon have their chance. In October 2019, Virgin Galactic unveiled the dark blue jumpsuit passengers will don, designed by Yohji Yamamoto. Of the uniform, Branson commented, "I think every single person who goes to space will be delighted with it . . . I think the whole experience of going to space should be sexy. Our spaceships are sexy. Our mother ships are sexy. Our spaceport is sexy. And for younger people than myself, this suit is also sexy."[37]

Just how sexy was Virgin Galactic? Rather like a World War Two fighter plane, the SS2 spaceship had painted on its nose an image of a young woman in a tight spacesuit arched back, with long blonde hair spilling out of her bubble helmet. On the opposite side of the ship's nose, the Virgin Galactic woman in the spacesuit is arched forward in a graceful swan dive. The pin-up style art becomes freighted with Freudian significance when you learn it is a tribute to Richard

Branson's mother, Evette, in her youth. (This makes more sense when you see her on the nose art of one of the WhiteKnightTwo "mother ships" named "Eve.")

Branson's Virgin Galactic was in a not-so-friendly rivalry with Blue Origin, owned by Jeff Bezos. Bezos, the world's wealthiest man in 2019, even though he has modeled his look after *Star Trek* Captain Jean-Luc Picard, has rarely been depicted as charming or roguish. One journalist described him as a "relentless," and "highly-motivated" master of predatory capitalism, dedicated to science fiction based ideas of progress.[38] Instead of "work hard, play hard," Blue Origin's motto, in stuffy Latin, is *Gradatim Ferociter*, "Step by step, ferociously." Instead of Galactic's cheesecake art, Blue Origin emblazons turtles on its spaceships, similar to the twinned turtles on its Renaissance-style corporate coat of arms. The turtles rear up from a sepia-toned map of the Earth flanking a shield as they look up to the blazing sun and planets. (Although he has styled himself after the turtle, Bezos's company was founded before SpaceX and Virgin Galactic.) Blue Origin will launch its suborbital flights from West Texas, as just one part of a wide business plan that includes satellite launches, a lunar lander, a potential lunar hotel, and a space satellite-based Internet system; for the long term Bezos has backed human emigration to orbiting space colonies.

One thing Branson and Bezos agreed on was ticket price. A seat on one of Blue Origin's suborbital flights on a New Shepard rocket will cost more than $200,000. Branson has said that he welcomed the rivalry, noting, "Both of us are going to do extremely well at it because the amount of people that want to go to space is enormous. We're not going to be able to build enough spaceships to satisfy demand." Branson also predicted that as more spaceships are built, the cost for a suborbital flight could come down to forty or fifty thousand dollars.

And for those who cannot afford even a $40,000 ticket and are unlikely to qualify as a NASA astronaut, there's always HERA missions (NASA wants you!), the brief Astronaut Training Experience at Kennedy Space Center, or a visit to the imaginary planet Batuu at Disney's Galaxy Edge. Me? I'm saving up for one of John Spencer's space yachts.

# CHAPTER EIGHT:

# THE MOON

• • •

"A lunatic . . . is one who . . . hath lucid intervals,
sometimes enjoying his senses and sometimes not, and that frequently
depending upon the changes of the moon."—William Blackstone, 1765

"One Acre of Moonland Standard Package: One-acre of moon land;
Moon land deed (name not included); Map with Selenographic coordinates."
—Groupon offer, February 1, 2020

A funny thing happened while I was writing this book—the attention of space enthusiasts shifted from Mars as the ultimate prize, to the Moon. This is what makes a historian queasy about the unfolding chaos of the present: its lack of fixity. The clear trend "Mars or Bust" emblazoned on my red tote bag the summer previous had morphed to: "we are going to the Moon (again, amen)." In October 2019, I visited the same physics conference room on the University of California campus that hosts the Experimental Cosmology Group, this time filled with undergraduates in sweatshirts and baseball caps (one read: "Deep Space") readying their entry for NASA's Big Idea Challenge 2020. While the 2019 challenge was to design a greenhouse for Mars, NASA's 2020 Challenge required students to develop tools to investigate the Moon's permanently shadowed polar craters for ice and other resources.

UCSB sophomore and physics major Anna Pedersen, with large glasses perched on her nose, was leading the student group in designing a pair of lunar rovers including a "mother" that could beam laser energy to recharge a roving "baby" unit that would descend into the lunar crater. The students were versed in space jargon such as: "payload limitations," "beamed energy," "PSRs (permanently shadowed regions)," "How can we get this to TR6?" (NASA's technical readiness level that indicates space qualified), the value of an "RTG" (radioisotope thermoelectric generator) for warmth, or a "Nano-stick spectrometer" (for identifying molecules).

Pedersen grew up near the Armstrong Flight Center at Edwards Air Force Base, an area where interest in spaceflight came naturally to those math and science oriented. As for NASA's renewed focus on the Moon, sitting cross-legged on a concrete bench after the meeting she said, "The Moon is interesting because as an opportunity it is more tangible. I can see it every night. Where I grew up, discussion of Mars research was commonplace. The Moon program was history. Almost forgotten. It would be great to deploy resources there."

"It seems more fresh?"

"Exactly."

Credit the subordination of Mars, in part, to the media frenzy in 2019 building to the fiftieth anniversary of the 1969 Apollo 11 landing. This media "shock and awe" campaign—which included the 2018 movie *First Man*, with Ryan Gosling starring as Neil Armstrong—softened feelings about the abandoned Moon, and suddenly, as Martian dust clouds faded, it became obvious that many corporations and nations all along had been Moon gazing: Japan, China, India, Israel, Russia, the European Space Agency, and the United States. The Moon was worthy.

U.S. President Donald Trump wanted NASA to pursue a Moon base (or was it an orbiting lunar station? A rest stop on the way to Mars?); in January 2019, China's Chang'e 4 rover landed on the far side of the Moon; in the spring, an Israeli lander bearing the message "Small Country, Big Dreams," crashed on the Mare Serenitatis; later that year the film *Ad Astra* was released, and in it, Brad Pitt's character visits a lunar colony that includes an Applebee's and a Subway sandwich outlet ... Basically, the interpretation of Buzz Aldrin's verdict on the Moon's "magnificent desolation" had shifted in emphasis from the second to the first word. Everyone, it seemed, wanted to get back to the Moon, and do it right, with their choice of bun.

• • •

I was born in the year of Sputnik, 1957, making me an unreliable if living witness to the emerging Space Age. Like other baby boomers, I treasured two moons: one from story books and the other that John F. Kennedy elevated into a questing object, all of its five billion cratered cubic miles, circling Earth in the vacuum of space.[1] I grew up reading, while sprawled on the carpet, Dick Tracey comic strips whose panels became puzzling when the alluring Moon Maiden from a crime-ridden Moon colony with her snail-like antennae showed up. In the basement, I made spaceships out of cardboard boxes using tinker toys as knobs and joysticks; I also hung spheres in shoeboxes to represent planets, and gawked at the wavy lines on a television screen when astronauts bounced slowly across the Moon's magnificent desolation. I doubt my family was the only one in the Midwest to carefully wrap in aluminum foil the July 21, 1969 edition of the *Chicago Sun Times,* with its headline "MEN WALK ON MOON," to preserve it for posterity.

Well before becoming a rallying point for American pride, an outer annex of Manifest Destiny, the Moon had a mixed reputation. For centuries it has been associated with dreams, love, poetry, the feminine principle, and the potentially sinister night world. In the novel *Ulysses*, James Joyce's protagonist Leopold Bloom, speculating on the "special affinities [that] appeared to him to exist between the moon and woman," offered a long reverie that included, "the forced invariability of her aspect: her indeterminate response to inaffirmative interrogation: her potency over effluent and refluent waters: her power to enamour, to mortify, to invest with beauty, to render insane, to incite to and aid delinquency: the tranquil inscrutability of her visage . . . her omens of tempest and calm: the stimulation of her light, her motion and her presence: the admonition of her craters, her arid seas, her silence: her splendor when visible: her attraction when invisible."[2]

In the Western tradition, the Moon has belonged to the goddess, and the presumed biologic link between women and the Moon is not merely the stuff of myth. The notion that menstrual periods synchronize precisely to the Moon's phases is incorrect, yet a worldwide average cycle of 29 days correlates well with the Moon's cycle of 29.5 days.[3] The word menstruate comes from the Latin root word "mensis" for "month," while the related "mene" in Greek, is one of the names of Selene, a pre-Olympian Moon goddess. Statues show her thick of body, with a smooth crescent horn above her brow.

Artemis (or Diana), the Greek Goddess of the Hunt, Chastity, and the Moon—before being forced on a diet like the torch-bearing Columbia Pictures logo lady—was also often depicted in this formidable way. Her rival hunters and/or would-be molesters (usually one and the same) came to bad ends. In ancient magic, Selene was regarded as one face of a triple goddess that included Persephone (Goddess of the Underworld), and Hecate (Goddess of Witchcraft). Before casting a spell, whether for good or harm, adepts heaped praise on the goddess and asked her help.

In addition to aiding magic, the Moon has been connected with madness, crime, "injurious dreams," "hallucinations," and, possibly worse—insomnia. Since the Moon influenced the tides, Aristotle and other ancients believed that moonlight disordered the workings of the brain, a notably moist organ. The extent of that disordering influence has continued to evoke concern. At least two modern medical researchers went on record to assert that "the incidence of crimes committed on full moon days was much higher than on all other days."[4] (Their findings have been disputed.)

Adding to its dubious reputation, in folklore and popular culture, the Moon, quite simply, produces monsters. Mooncalves, in Shakespeare's time, injured by exposure to moonlight, were the deformed fetuses of farm animals and humans. A few centuries later, on silver screens, moonbeams cued vampires to rise from crypts—and when wolfsbane bloomed, the orb's fullness turned Lon Chaney Jr. into a reluctant, but highly effective monster.

The Moon's eerie reputation also informed scientific romances. Astronomer Johannes Kepler, deviser of the three laws of planetary motion, set the tone with his narrative *Somnium*, written in 1608, in which daemons first drug with opiates, and then, with explosive force, carry hearty Earth dwellers to the Moon. The selection process, like that of NASA's astronaut program, was rigorous, "We do not admit sedentary, corpulent or fastidious men into this retinue. We choose rather those who spend their time persistently riding swift horses or who frequently sail to the Indies, accustomed to subsist on twice-baked bread, garlic, dried fish, and other unsavory dishes." The ideal candidates, the ones with the right stuff, according to the daemons, were "dried up old ladies . . . [who] from early childhood are accustomed to riding goats . . . and to travel through narrow passes and through the immense expanse of the Earth."[5] The image of a woman riding on a goat comes from seventeenth-century depictions of the witches' Sabbath in which the devil is said to arrive on a goat, with said goat prominent in the ensuing orgy.[6]

According to Kepler, on the Moon all births are monstrous, and plants and forests grow rapidly during the long Moon daytime (about 14.75 Earth days). Kepler's Moon dwellers grow enormous, travel in hordes, and die soon, while chasing the receding waters that boil at the surface and freeze in the depths. Those boiled alive are eaten. All lunar species, plants and creatures, have a thick "spongy and porous" skin that the Sun may scorch and cause to fall off. Many Moon dwellers take to underground caves to avoid the extremes of heat and cold.

Kepler's account became the template for H. G. Wells's *The First Men in the Moon* (1901) written three centuries later. Bedford, the failed playwright and rogue who narrates this satire, joins the scientist Cavor on a spaceship with a supply of a miraculous antigravity substance that propels them to the Moon. Soon after landing, they observe mushroom forests rapidly appear. These forests serve as fodder for giant mooncalves herded about by ant-like humanoids who eventually retreat to their high-tech civilization underground. Bedford escapes with booty, and on the flight home experiences an ego-shattering

Overview Effect, but perhaps it doesn't "take," because he immediately returns to his greedy ways upon landing in England.

• • •

Journeys to the Moon remained in the realm of fantasy even after Tsiolkovsky wrote his paper in 1898, "The Exploration of Cosmic Space by Means of Reactive Devices." President William McKinley, for example, was said to have enjoyed a ride on a simulated "Journey to the Moon"—a ship complete with flapping bat-like wings—before his assassination at the 1901 Buffalo world's fair. In his 1919 article "A Method of Reaching Extreme Altitudes," Robert Goddard avoided discussion of manned spaceflight but casually dropped the idea of sending flash powder via rocket to a new Moon ("2.67 lb to be just visible, 13.82 lb or less to be strikingly visible"). Ignited on impact, the flash would be seen from Earth with telescopes.[7] This was sensationalist enough to get him hounded by the press. By 1957, when the Soviets launched Sputnik, talk of Moon flights no longer indicated the speaker had an overly-moist or disordered mind. The race was on.

John F. Kennedy wanted to out-do the Soviet space program, which kept achieving "firsts"—first satellite, first animals in orbit, first lunar probe, first man in space, first probe to Venus—soon it would be the first woman in space, first spacewalk, and so on. No doubt about it, based on the heft of the satellites the Soviets launched, their rockets were bigger than ours. In April 1961, Kennedy queried Vice President Lyndon Johnson, "Do we have a chance of beating the Soviets by putting a laboratory in space, or by a trip around the moon, or by a rocket to land on the moon, or by a rocket to go to the moon and back with a man. Is there any other space program which promises dramatic results in which we could win?"[8]

When Johnson asked Wernher von Braun for his assessment, von Braun responded that the United States would not likely beat the Soviet Union in launching a laboratory into space, and only had a "sporting chance" of landing an unmanned probe on the Moon or sending a manned crew in orbit around the Moon before the Soviets. However, von Braun suggested there was an "excellent chance" the United States could beat the Soviets to a manned landing on the Moon, as neither the Soviets nor the United States currently had sufficient rocket power, and by marshaling resources and treating it on par with a "national emergency," the Moon might be reached by 1967 or 1968.[9] Johnson told Kennedy, "Manned exploration of the moon . . . is not only an achievement with great propaganda value, but it is essential as an objective whether or not we are first in its accomplishment—and we may be able to be first."[10]

Von Braun's timetable proved largely accurate. In late May 1961, Kennedy announced to Congress the goal of sending a man to the Moon by the close of the decade, outlined a budget of $7 to $9 billion, and added, "Money alone will not do the job. This decision demands a major national commitment of scientific and technical manpower, material, and facilities . . . [so] this nation will move forward, with the full speed of freedom, in the exciting adventure of space."[11] Not liquid oxygen or hydrogen, but "freedom" would be the fuel to open outer space.

On September 12, 1962, Kennedy gave his well-known speech at Rice University announcing the push to reach the Moon. He focused on patriotism and the national good while indicating that beyond patriotism were the larger goals of humanity. Aware of the 1961 U.N. resolution that called for "international cooperation in the peaceful uses of outer space," he commented, "There is no strife, no prejudice, no national conflict in outer space as yet. Its hazards are hostile to us all. Its conquest deserves the best of all mankind, and its opportunity for peaceful cooperation may never come again." He added, "We choose to go to the moon in this decade and do the other things, not because they are easy, but because they are hard, because that goal will serve to organize and measure the best of our energies and skills, because that challenge is one that we are willing to accept, one we are unwilling to postpone, and one which we intend to win."

Von Braun's original proposal involved an enormous multistage rocket, the Nova, capable of landing on the Moon and returning. Alternate plans included building a space station in Earth orbit where

Moon shuttles could be assembled, or, to keep costs down, to send two or three Saturn rockets into Earth orbit, the additional rockets to refuel a ship that could land on the Moon and return to Earth. Von Braun eventually approved the approach Apollo 11 relied on: a lunar orbit rendezvous, in which a small lunar module would separate from the main ship to taxi to the Moon's surface and back. Only one Saturn rocket was required.[12] If nothing else, this risky but economical plan ended the speculation offered in Tom Wolfe's *The Right Stuff*, that astronauts, unlike the active Air Force test pilots from whose ranks they were drawn, were simply passengers in tin cans, passive agents who had to beg for windows and controls in the early space capsules. As it turned out, neither Neil Armstrong nor Buzz Aldrin were able to simply sit back and enjoy the ride to Tranquility Base.

With Armstrong at its controls and Aldrin behind a duplicate set, reporting altitude and speed of descent, on July 20, 1969, Apollo 11's LEM (Lunar Module) separated from the Command Module. Nicknamed the Spider, the LEM was an all-but untested, spindly, gold foil-shrouded device with four landing legs. Approaching the Moon, it had no atmosphere to contend with or to help in braking. It began its descent at an opening speed of 3,600 miles per hour, tilting entirely upside down at 45,000 feet above the Moon. Somewhere below 40,000 feet, the overtaxed onboard computer began to flash a warning. Mission control gave the astronauts the "Go" command to override. By the time they reached 13,500 feet, the LEM had slowed to an earth-bound passenger jet speed of 600 mph. The astronauts continued, not spotting the expected landmarks on the lunar surface, ignored another computer warning, and at 1,000 feet, with fuel dwindling, approached a large boulder-strewn crater floor. Armstrong recalled, "The rocks seemed to be coming up at us awfully fast." They were nearly out of fuel when the LEM cleared the crater wall and Armstrong picked out a landing spot "the size of a big house lot." With radio bursts serving as quotation marks, Armstrong noted "The Eagle has landed."[13]

When the astronauts returned to Earth, President Richard M. Nixon exclaimed, "This is the greatest week in the history of the world since Creation." All looked grand for a continued and vigorous human space program. How would it take shape? Von Braun brought out his 1950s laundry list: a space station, a space shuttle, a permanent Moon base, and an excursion to Mars by 1981.

As for how a permanent Moon base might shape up—during the Eisenhower presidency, the U.S. Army had developed a Moon base dubbed "Project Horizon." Horizon, whose design von Braun had assigned to Heinz-Hermann Koelle, a Paperclip colleague, had as its goal a scientific—and military—outpost for twelve men on the Moon by 1965. Excavators would dig trenches on the Moon for ten 20-foot-long and 10-foot-diameter cylinders. These would be covered with three feet of lunar material to protect personnel from radiation. The interlinked cylinders would provide housing, along with modules for dining and recreation, communications, medical facilities, and toilets.[14] When leaving the underground base through airlocks, personnel would don two hundred pound spacesuits.

A small nuclear reactor would provide heat and tanks of oxygen and nitrogen would provide atmosphere. Food, initially, would be pre-packaged in paste form, "however, as water supplies increase with the introduction of a reclaiming system, dehydrated and fresh-frozen foods will be used."[15] The project designers also insisted algae might prove a palatable food source. A radio relay station would connect with three satellites in Earth orbit. To make the outpost habitable would require about 150 rocket launches, with five or six launches a month. Another 64 launches yearly would keep the post running. The army put a six billion dollar price tag on the project.

The two-volume report made at least one reference to caching weapons in supply areas, however it did not detail, as Internet lore has it, that a Claymore anti-personnel minefield planted around the base would keep out "hostile forces" or that tactical nuclear weapons would be available to destroy enemy vehicles. But the report did note that in addition to military surveillance capabilities, the base might eventually house nuclear missiles targeting the Earth.

Horizon's developers agreed with the House Select Committee on

Astronautics and Space Exploration's assessment that "The military potentialities of space technology, which the United States would prefer to see channeled to peaceful purposes, are greater than general public discussion to date suggests." In a close approximation of doublespeak, the House report added, "Military space capabilities are technically inseparable from peaceful capabilities which are well worth pursuing in their own right." The report underlined that space-based reconnaissance to patrol "merchant ship lanes" could also be used to pinpoint bombing sites, while communications "to improve global relations" could be used for "controlling military forces." More chilling, "Rockets for cargo and passenger delivery can also carry thermonuclear weapons," while "satellites designed to return men from orbit . . . can also deliver bombs."[16]

Dwight Eisenhower wanted nothing to do with Project Horizon. Milton Rosen, NASA's Director of Launch Vehicles recalled Eisenhower's response in 1959, "While people in the room thought the briefing was stimulating, President Eisenhower was very negative." He did not want to be remembered as someone who would encourage dubious missions to the Moon.[17]

Project Horizon was one of the first serious proposals for a Moon base. At the same time the army was planning Horizon, the Martin Company (later Lockheed-Martin) floated plans for a simulated "moon house" on Earth, consisting of a pressurized steel sphere with a diameter of thirty-two feet inside another sphere that maintained a vacuum. Air locks would lead subjects to the inner sphere that housed mechanical systems, crew quarters, a kitchen, recreation room, library, a small science lab, as well as a hydroponic farm to grow algae and replenish oxygen and another for vegetables.[18] Martin Company also developed a lengthy report on "Lunar Basing" for a 1961 symposium that included alternatives for above-ground, below ground, and inflatable shelters.

In addition to military and scientific value, the company promoted the Moon base's potential as a tourist destination and health sanitarium. Regarding the therapeutic properties of the lunar environment,

Dandridge Cole, one of Martin Company's rocket designers, in 1960 argued that the Moon's low-gravity could benefit arthritis patients, as well as "paralytics and other cripples. It could be helpful for heart patients as well." Indeed, he added "there is a strong possibility that normal life span would be increased by living under conditions of a lunar colony."[19]

A Moon base, then, could combine health spa, scientific outpost, and nuclear missile launch site. Wernher von Braun had long promoted the idea of a nuclear-armed space outpost, and the idea intrigued some military officials. In a 1959 interview, Air Force Brigadier General Homer A. Boushey, stated, "We definitely recognize the probability of military operations on or in the vicinity of the moon . . . We ought to consider the possibility of moon-based weapons systems, eventually to be used against earth and space targets." As it would take less energy to launch warheads from the Moon to the Earth than from the Earth to Moon, Boushey commented, "From an energy viewpoint, the moon represents the age-old military advantage of high ground."[20] He also thought the Moon could serve as a "retaliation base of unequaled advantage," if the Soviet Union initiated a missile attack.

Through most of the 1950s, von Braun pressed this perspective, insisting that space stations or Moon bases should not only be used for surveillance but be armed with nuclear missiles. However, after he joined NASA as director of the Marshall Space Flight Center, he moderated his views. Otto Binder, a science fiction writer and editor of *Space World* approached von Braun in 1961 for a planned article on "Atomic War from Space." To questions such as: "Will space warfare be worse than ICBMs?" or "Can space become a warfare arena in the future?" von Braun answered, "In view of the fact that NASA is engaged in peaceful exploration of outer space, I believe it would be inappropriate for me to comment on military subjects like those set forth in your questions. I must, therefore, respectfully decline."[21]

The concept of "peaceful cooperation" in outer space was gaining momentum. The United Nations did not want nuclear weapons in Earth orbit or on the Moon. Not only had a "Rockets for Peace" campaign begun in 1955 as part of the build-up to the National Geophysical Year of 1957, but the U.N. was laying the groundwork for what would eventually be the Outer Space Treaty, ratified in 1967, which, among its resolutions, noted, "The Moon and other celestial bodies shall be used exclusively for peaceful purposes."

Well before the signing of the treaty, reported plans of the United States to turn the Moon into a nuclear missile base did not faze Soviet officials. One AP article, "Red Says Moon Won't Be Armed," quoted the Soviet scientist Z. K. Fedorov: "No one in the U.S.S.R. is planning to fence off the region of the moon where the Soviet pennant flies and start building military bases . . . Perhaps this will have a certain soothing effect on certain ardent Western military men who . . . daydream aloud how wonderful it would be to threaten whole nations by shooting them from the moon."[22] According to historian Asif Siddiqi, there was a Soviet proposal in 1958 to explode a nuclear device on the Moon. Similar to Goddard's flash powder experiment, the goal was, Siddiqi noted, "to see if it was possible to observe the explosion from the Earth. This proposal was put forth by the Soviet physicist Yakov Zeldovich but was never considered past the proposal stage."[23] The U.S. Air Force also considered, and then rejected, a similar bombing mission.[24] The Soviets settled for landing the Luna-2 probe on the Moon in 1959.

While they didn't appear to regard the Moon as the ultimate military outpost, the Soviets did insist that it would be there that, in 1967, they would celebrate the fiftieth anniversary of the Bolshevik Revolution. In an undated wire story "Soviet Moon City Planning" a Soviet official described a future "Red city" on the Moon, relying on both solar and nuclear power. The city would be inside a crater "under a tremendous dome of glass and supplied with aluminum doors or 'air-locks.' Inside, at every half-mile or so, the city will be partitioned by glass walls with double doors. These will minimize the damage that may result from falling meteorites or other accidents."[25]

Following the triumphant landing of Apollo 11 in 1969, a Moon base was only one component of NASA's grand post-Apollo vision.

Mars remained a grander prize. But von Braun's premise of pressing forward with "national emergency" levels of funding did not catch on. NASA pushed von Braun's entire shopping list, including the manned Mars expedition, but Nixon's team, pulling an Eisenhower, decided to only fund the space station and shuttle.

• • •

Although Afrofuturist big band jazz leader Sun Ra often punctuated concerts with the chant "Space is the place," the government's spending spree on the Apollo program had always been controversial, particularly among Black Americans. The year prior to the launch was marked by Martin Luther King's assassination and urban uprisings. The day before the Apollo 11 launch, Ralph Abernathy, one of the organizers of the Poor People's Campaign, led a march, complete with a wagon pulled by mules, to the Cape Canaveral launch site in Florida and commented, "A society that can resolve to conquer space . . . deserves acclaim for achievement and contempt for bizarre social values . . . Why is it less exciting to the human spirit to enlarge man by making him brother to his fellow man? There is more distance between the races of man than between the moon and the earth."[26]

While a crowd gathered in Central Park to watch the landing on giant television screens, in Harlem a crowd of 50,000 at a soul music concert booed when the landing was mentioned. The next day, the New York Amsterdam News offered, "Yesterday the Moon, tomorrow maybe us."[27] One year later a writer in Ebony noted, "From Harlem to Watts, the first moon landing in July of last year was viewed cynically as one small step for 'The Man,' and probably a giant leap in the wrong direction for mankind. Large segments of the rest of the population, except perhaps at the time of the first landing, were merely bored."[28]

And it was not just America's minority population that had tired of the landings. While the missions ground on, for television viewers back on Earth, the Apollo program had turned stale. When President

Richard Nixon officially ended the Apollo program in 1972, most Americans were in accord. A 1970 Harris survey indicated a majority (56 percent) did not think the landing was worth its price tag. This was, however, greater support than in 1967, before the landing, when only 33 percent of the public thought the money spent on space exploration was worth it.[29] We had collected plenty of rocks. Left a flag. Determined the Moon to be devoid of life. We seemed done, perhaps for good, with the Moon.

Despite this growing apathy, space diehards soldiered on, particularly as the hobbled agenda of the space shuttle program became clearer. The do-it-yourself approach, expressed earlier in the founding of rocketry clubs, was taken up post-Apollo by enthusiasts who felt betrayed by the retreat from manned spaceflight. While many true believers gravitated to Gerard O'Neill's space colony idea, others contemplated colonies on the Moon, Mars, and asteroids. Even O'Neill noted that a lunar outpost would be needed to strip materials from which to build Lagrange colonies. Others realized that if your goal was space settlement you could do worse than start with the Moon itself. Digging one meter down, temperatures are a constant -7 degrees Fahrenheit, while twenty meters underground a habitat could be created with "temperatures like an earthly home."[30] (The heat source: a partially molten zone surrounds the Moon's iron core.)

The longtime commander-in-chief of the D.I.Y. Moon community during this dry spell for manned flight was Peter Kokh of Milwaukee. Born in the 1930s, a seminary student turned industrial worker and repairman, Kokh began researching a science fiction book set on the Moon and ended up captivated by the possibility of Moon settlement. In 1974, he was one of the first to join the National Space Institute, the advocacy group that Wernher von Braun founded after leaving NASA. From 1986 to 2012, Kokh also edited and wrote most of the copy of *The Moon Miners' Manifesto* (ten issues per year). The Moon Society, which began in 2000, with Kokh as its second president, adopted the *Manifesto* as its official publication. The magazine's motto was "expanding the human economy through off-planet resources." For decades, Kokh, an equal opportunity space enthusiast, proselytized about settling the Moon, L-5 sites, Mars, and asteroids. Of a possible moonbase, Kokh rejected the Project Horizon vision, writing, "It is not enough to provide a shelter for scientists so that they can titillate their curiosity, and ours, and then return home. Our purpose must go well beyond this to make the Moon *a second human world.*"[31]

Kokh's *Moon Miners' Manifesto* offered what might be called the *Whole Moon Catalog* for lunar homesteaders. Themes and topics repeated: viable construction processes using local ("in situ") resources, industries and crafts necessary to build and sustain a colony, ways to create an authentic Moon culture, the importance of rugged individualism, the dangers awaiting the unprepared, how settlers on the Moon and Mars must become trading partners. But you never knew exactly what you might find. One article promised that despite limitations on paper on the Moon, toilet paper could be recycled for those who deemed it a necessity; new funeral practices might emerge such as a "desicatorium" where bodies could be freeze-dried; he devised a Moon calendar based on the unit of a "Sunth," the sunrise/sunrise schedule of (29.5 Earth days); he also promoted Moon friendly sports such as handball, racquetball, jai alai, skiing (down the frozen edges of craters), and spherical trampolining.

Kokh reminded readers of the dangers of sharp-edged Moon dust; proposed the practical value of dwarves or little people as spaceship crews; promoted the social role of taverns on the Moon (with names such as "The Long Dawn" or "The Dusty Boot"); offered ruminations on developing an indigenous Moon music (he suspected a gamelan-like orchestra of bells and gongs might be the most appropriate with home-grown instrumentation); and reflected on "the spiritual aspects of settlement." What started as research for a science fiction novel, ended up as an encyclopedic study of Moon settlement.

Although likely not directly inspired by the *Moon Miners' Manifesto*, many of Kokh's speculations enlivened the pages of subsequent science fiction. *Manifesto* readers were especially keen on the idea of putting settlements inside the Moon's large underground "lava tubes," stable structures for billions of years, which are shielded from radiation and offer a near constant temperature of -4 degrees Fahrenheit. Arlin Crotts, an astronomer at Columbia University has noted that lava tubes, potentially, could offer "living space for millions." He added that "tubes are so large that one can engineer spaces sufficient to engender feelings of being outside."[32] (Kim Stanley Robinson's *Red Moon* features a lava-tube colony, the secret project of a Chinese billionaire. Modeled on Tang Dynasty paintings of an idealized countryside, it includes a river stocked with fish, house boats, pagodas, peach trees, and birds flying below artificial clouds. Ian McDonald's *Luna* series also features clans who create biomes in lava tubes—"the rock trembles to the vibrations of excavators and sinterers working deep in the walls, shaping rooms and spaces for Luna and her generations."[33])

Several issues of the *Moon Miners' Manifesto* took up the idea of settling "rilles" or large channels on the Moon—up to several kilometers wide, and hundreds of kilometers long—many possibly formed from collapsed lava tubes. Kokh and collaborators devised an enormous rille-bottom settlement they named "Prinzton." Its framed roof, secured with cables, would be covered with lunar regolith as radiation shielding. An upper layer of the settlement would be used for farming, with heliostats directing in sunlight; a second roof would cover three villages, including homes, parks, and transportation. Kokh answered the question why, with the retort, "rilles have sides! . . . a rille is an excavated foundation just waiting for construction!"[34]

Construction methods could include the use of sintering—heating, in this case, lunar regolith to form a glassy surface. NASA experts, often proponents of dropping modular habitats on the Moon, also have taken to the notion of using "in situ" resources. The authors of a 2005 article in an aerospace engineering journal determined that if sintered or fused into glass by microwave, lunar regolith could effectively hold an atmosphere. The authors experimented with both lunar "simulant" and actual Moon soil samples (eighty grams from Apollo 17) and found that, whether in making bricks, roadways, underground habitats, or building material, "The applications that can be made of the microwave treatment of lunar soil for in situ resource utilization on the Moon are unlimited."[35]

One of the most significant contributions of the Moon Society along with the National Space Society was to advocate for the Lunar Prospector. Kokh helped organize meetings in 1988 and 1989 to advocate for a probe to orbit the Moon's poles and map its surface, and search in particular for ice in craters. The project gained the interest of NASA scientist Alan Binder. "I came up with the name Prospector and it stuck," Kokh commented. NASA's Prospector mission, launched a decade later in 1998, established that there was not only ice at the poles but in many permanently shaded craters near the poles.

When I asked Kokh, then retired in 2019, about current enthusiasm for the Moon, he answered, "Most people are more interested in Mars. Most people, regarding the Moon, see a dry gray desert." He chuckled and added, "Scientists want to go to either the South or North Pole because of the ice craters. They want the ice for fuel. I take a totally different view. It started when I found out what you can make with basalt [volcanic rock] common in the Moon's *maria*—gray areas—you can cast basalt, you can make it into fiber. You can make habitats with it, furnishings, and so on." He chose one large area to the north, Mare Frigoris, and said he would put a settlement at one end of the mare. "You'd get ice from the craters. If you put another settlement on the far end of the mare, with a cable between, when dawn rose everyone could have [solar] electricity for three out of four weeks, instead of just two."[36]

• • •

One of the more recent high profile science fiction books set on the Moon took few cues from the *Moon Miner's Manifesto*. Eschewing lava tubes and rilles, Andy Weir's *Artemis* (2017) presents a thriving settlement on the Moon that includes large domed areas, superficially reminiscent of shopping malls, named after Apollo astronauts Bean, Conrad, Aldrin, Armstrong, and Shepard (he of the mighty golf swing).

The settlement also extends at least fifteen stories below ground. A train from Artemis takes visitors forty kilometers to the Apollo 11 Visitor Center on the Sea of Tranquility, where, through glass walls, tourists can view the bootprints left behind by Armstrong and Aldrin, and the remains of the Apollo 11 descent module. Tourists also take guided moonwalks of the site, which is treated with reverence by visitors and settlers alike.

I asked Weir why he did not employ lava tubes as a colony site in *Artemis*. He replied that lava tubes are great for radiation protection, but added, "I don't like that plan for a few reasons. First off, and most importantly, it severely limits the locations that you can place your settlements. Those lava tubes aren't all that common. To me, that's a deal-breaker. Especially in the early days of lunar settlement which, I believe, will be about tourism. And tourists will want to see Apollo landing sites. This means your settlement has to be near the Apollo 11 site. And there are no lava tubes anywhere near there."[37]

Realistically, how might Moon settlement unfold? Weir answered, "I think the first human settlement on the moon will be tourism-based. And that can only happen when middle-class people can afford to go there. But that's just a matter of technology." Considering that he wrote his Mars novel first, was he in the Moon-first or Mars-first camp? He answered: "Moon first, all the way. Colonizing Mars before colonizing the Moon would be like if the ancient Britons colonized North America before colonizing Wales." He admitted to having a soft spot for Mars, noting it was "ideal in the long term, because it has all the elements necessary for a biosphere to grow. There is carbon, hydrogen, oxygen, and nitrogen in great abundance."

• • •

For decades after the Apollo program ended, the likelihood that the United States would send another human to the Moon was low. Post space shuttle, NASA's focus was on robotic exploration of the solar system. Few, outside the several hundred members of the Moon Society and the roughly 10,000 members of the National Space Society were particularly interested in new moonwalks. But this isn't to say that administrations didn't try to drum up business for NASA and its aerospace contractors. George W. Bush backed NASA's $100 billion "Project Constellation," which promised a Moon base by 2025, as part of an eventual Mars program. In 2009, budget experts deemed $100 billion an underestimate of its potential costs.[38] Newly elected President Barack Obama pulled an Eisenhower.

Prospects changed in 2016, when the Trump administration not only championed a new branch of the military, the Space Force, but also NASA's Gateway program. This program included: (1) Gateway, a small space station in lunar orbit; (2) the use of the not-yet-completed Space Launch System (SLS) rocket and Lockheed Martin's Orion spacecraft to shuttle from Earth to the Gateway; (3) smaller landers to travel to the Moon's surface and back to Gateway, and (4) eventual deep space shuttles from Gateway to Mars orbit and landings. The moon landing project was named Artemis, the Greek name for the Moon Goddess Diana, Apollo's twin. The likely landing place, as Kokh predicted, would be near the Moon's South Pole and one goal would be studying water resources. NASA administrators made it clear they wanted a woman to be the next to add bootprints to the lunar surface—all twelve Apollo astronauts who walked on the Moon were white American males. Janet Kavandi at NASA's Glenn Research Center and a former astronaut said, "I think most women would say it's about time."[39]

The Gateway plan included placing a space station in a high lunar orbit (about 600 miles above the surface), and sending landers down to the Moon by 2024—or 2028, and shuttles to Mars by the 2030s or 2040s.[40] The scheme was elaborate and costly—NASA estimated thirty-seven launches would be necessary to assemble Gateway. The Space Launch System (SLS) rocket whose contractors include Lockheed Martin, Boeing, and ATK was still not complete as of 2019. (The SLS was to carry the Orion spacecraft that would rendezvous with Gateway.) When and if in operation, the SLS would be about 15 percent more powerful than the Saturn rockets that brought Apollo to the Moon. Unlike rockets built by SpaceX and Blue Origin, the SLS had no reusable stages; flight costs were estimated to be eleven times that of SpaceX's Falcon Heavy Rocket. As of 2019, production of the SLS and the Orion spacecraft had already cost NASA $50 billion.

Seeking to shift public—and institutional—interest from bold Mars colonizing plans back to the Moon, NASA officials resembled harried parents urging their children to eat their waxed beans and tie their shoes—*and no faces, young lady*. In 2018, NASA Administrator Jim Bridenstine said, "We need to get to the moon in order to learn how to live and work on another world so that we can go to Mars . . . The moon is the proving ground, Mars is the destination."[41] The Moon was convenient, and was a cheaper and safer place to test systems for living off-planet. (A journey to the Moon takes about three days, while Mars requires a minimum journey of five months one-way and a long wait for a viable return.) Echoing Bridenstine, NASA planetary scientist Chris McKay, a longtime proponent of Mars terraforming, commented, "My interest is not the Moon. To me the Moon is as dull as a ball of concrete. But we're not going to have a research base on Mars until we can learn how to do it on the Moon first. The Moon provides a blueprint to Mars."[42]

Few in the space community or wider public appeared dazzled with Gateway. In a 2019 Ipsos poll, only 9 percent of Americans believed returning to the Moon should be a top priority in space. Robert Zubrin, a frequent critic of NASA's big money, bureaucratic approach to space, dubbed Gateway "Project Tollbooth"—a way to keep money flowing to NASA contractors such as Lockheed and Boeing. Zubrin proposed, instead, a Moon Direct plan. He argued there was no need for a lunar orbiting station, but rather a base on the Moon where fuel, eventually, could be manufactured from in situ resources for further flights. Lunar excursion vehicles could shuttle between the Moon and the ISS.

In sync with Zubrin, Apollo astronaut Buzz Aldrin also suggested a better staging ground for the Moon or Mars would be a space station in low Earth orbit. This station could easily be reached by existing rockets built by SpaceX, Blue Origin, and others—and could also replace the ISS when it is decommissioned, likely in 2030 or earlier. Aldrin noted, "Commercial launch vehicles and those under development, that are reusable, can handle pretty much everything."[43]

Former NASA head Michael Griffin concurred—to a degree. He argued it was better to formulate a Moon direct plan and begin to process rocket fuel from water harvested on the lunar surface. Griffin remarked, of the lunar orbiting station, "Gateway is useful when, but not before, they're manufacturing propellant on the moon and shipping it up to a depot in lunar orbit."[44] Novelist Weir agreed. "I like the idea of a permanent presence on the moon. Scientists working there all the time—it seems like a logical next step after ISS. I'm not too excited by Gateway. That seems like a needless extra step."[45]

Dale Skran, a former Bell Labs engineer and an official with the National Space Society, commented, "My view is that we need to first define why we are going back to the Moon. To mine water for rocket fuel? To build a radio telescope on the far side of the Moon? Only then should we develop a plan for going back. The Artemis program seems focused on a political stunt of landing the first woman on the Moon, with no commitment to anything beyond that. I want to see the U.S. land a woman on the Moon, but I don't want her to walk around, collect a few rocks, and come back to Earth as the political gas goes out of the program."

As for Gateway, Skran noted, "Personally, I am not as opposed to it as Zubrin, but I think it is important that NASA expand the scope of their architecture to fully include commercial vehicles . . . It would be far better to start with a clean sheet of paper, define a sustainable, low-cost architecture for supporting a base on the lunar surface, and then see how SLS/Orion might fit in . . . In the long run I expect there will be a Lunar Gateway, but it may be quite different than currently envisioned by NASA."[46]

His analysis proved accurate. In March 2020, Doug Loverro, NASA's associate administrator for human exploration and operations, announced that just as critics had urged, Gateway would be sidelined.

With the Trump administration pressing for a 2024 Moon landing, Loverro insisted that the Gateway plan was too ambitious and had too many untested aspects to rush through safely. Gone but not forgotten, Gateway would be reserved for future "sustainable" missions."[47] When would Gateway be built? Unclear. What would shape up instead? While Robert Zubrin favored his Moon Direct plan, he suspected we would instead get what Skran had critiqued—an updated version of Apollo 11. Instead of a Saturn rocket there would be the SLS, while Orion would take the place of the Command Module, along with a simplified lander along the lines of the LEM.[48] Whether a permanent Moon base would be established also was not clear, but the NASA announcement suggested not in the short term. The entire 2024 mission could become, largely, "a political stunt."

Establishing the Moon as a proving ground had become secondary, leading to the question: why go at all? Eighty-six percent of Americans polled in 2018 thought sending astronauts again to the Moon was either a low priority or entirely unnecessary. They were not alone. Of the cultural and historical significance of a return to the Moon, science writer Oliver Morton has argued that though it once epitomized a noble high-tech future, "Now it seems, at best, a future among others—and a slightly retro one."[49] Back in 2009, Apollo astronaut Buzz Aldrin noted, "There's not a more desolate location that I have ever seen . . . it's not a good place to set up housekeeping."[50]

The simplest answer to "why the Moon?" seems to be that space 2.0 is heating up and the Moon is the easiest target in the sky. Many U.S. space advocates hope that competition with China, with its ambitious space program and plans for exploring both the Moon and Mars, will spur American spending and interest. China will likely launch taikonauts to the Moon by the late 2020s. Russia also has plans to land cosmonauts on the Moon by 2030. This nationalistic narrative of the new "race" to the Moon suggests an unwelcome return to the days of Cold War posturing.

Even in science fiction, Moon settlement tends to the dystopic. Such outposts usually double as penal colonies, settings for dramas of suffering and longed-for liberation. Kim Stanley Robinson's *Red Moon* includes brief visits to a utopian community in a "Free Crater" whose multinational inhabitants run a quantum computer, mine ice, and swing about on ropes like aerialists during daily life and collaborative performances, but beyond the gesture of carrying a female astronaut to the Moon, it is hard to sense a utopian "vibe" from Artemis or even Gateway. Absent from Gateway discussion was Peter Kokh's vision that creating an economic base is only a step toward generating a vibrant new culture.

The private sector, of course, has other goals. Key to the current race to the Moon is not so much national prestige as potential profit from mining. One obstacle is the Outer Space Treaty of 1967, which notes: *"The exploration and use of outer space, including the moon and other celestial bodies, shall be carried out for the benefit and in the interests of all countries, irrespective of their degree of economic or scientific development, and shall be the province of all mankind."* The space treaty also states that outer space is *"not subject to national appropriation by claim of sovereignty,"* and that individual nations must assure that non-governmental entities under their jurisdiction abide by the treaty.

The treaty seems to have been only a temporary roadblock to corporations and their legal teams. Ignoring the treaty's intent, in 2015 U.S. President Barack Obama signed into law a statute permitting U.S. citizens to legally own resources mined in outer space. The small European nation Luxembourg, a banking powerhouse eager to secure the business of New Space corporations, followed suit in 2017. Other countries are contemplating similar laws.

The initial target for lunar prospectors will be water for rocket fuel and life support, offering high value in outer space. Next, various international groups are proposing to mine Helium-3, to be used in fusion reactors, assuming viable fusion technology emerges. Helium-3 is 100 million times more abundant on the Moon than the Earth. (Fans of the 2009 movie *Moon* will recall this was the product that the movie's protagonist, a radiation-addled clone, was mining along with robotic helpers.) Lunar mining advocates are also interested in the rare earth metals needed in electronic products such as cell phones, wind turbines, and electric car batteries. Ninety percent of rare earth metals now come from China, where authorities report the supply may be exhausted by 2035. Orbital sensors have indicated the presence of these rare earth minerals on the Moon, but little more is certain.

Some groups are intervening with proposals to avoid the Moon becoming a site of the feral capitalism depicted in Ian MacDonald's *Luna* science fiction series. The European Space Agency (ESA) for example, is supporting a "Moon Village" scheme that would make development of the Moon similar to the International Space Station. The project's somewhat vague goal, according to ESA Director General Johann-Dietrich Woerner is to "create an environment where both international cooperation and the commercialisation of space can thrive." The Village might include participants from industry, science, and tourism. Villagers would "leave behind on Earth any differences of opinion."[51]

The architectural firm Skidmore, Owings, & Merrill and M.I.T. have begun a master plan for ESA's Moon Village, to be sited near Shackleton Crater (on the south pole), with modular structures that will eventually give way to buildings made of local materials, connected with tunnels. Heading the design team at Skidmore, Owings, & Merrill, architect Colin Koop agreed that ESA's goal of "thinking beyond the boundary of a single country was attractive." While their focus for the moment was to prove feasibility, if and when that was accomplished, "we will shift to find ways to grow the collaboration and fabricate and test components."

Another member of the design team, Daniel Innocente added, "Sometimes you need a target to encourage capabilities, as with the Apollo program. A vision gives people ideas of what is possible if we work together." As to what differentiated their plan from numerous other lunar base models, Koop said that because of lower launch and payload costs, and a new line of rockets with greater carrying capacity, a near term lunar base was feasible. They offered one that could be adapted to different uses. While the main goal remained to keep people alive, architects could help make long term living off-planet psychologically satisfying. "Beyond creating pretty renderings we can bring a clinical approach to ensuring well-being. That's where design has a role."[52]

In addition to ESA, the Open Lunar Foundation—a group that includes Silicon Valley executives and former NASA leaders—also proposes to forge an "open source" lunar settlement. Their settlement would not be indebted to any one nation or corporation but function much like the ESA's Moon Village. All of this sounds better than imaginary lunar colonies surrounded by landmines and soldiers with tactical nuclear weapons, but until Open Lunar raises three billion dollars, it remains on the utopian drawing board.

• • •

Even in ancient myths, the Moon was a place of exile. While the story of Chang'e, the Chinese Moon goddess, has variations, in perhaps the most family-friendly, she guards an elixir of immortality that her husband, Houyi, an archer, had won, after shooting down nine suns that were scorching the Earth. When the archer's apprentice attempts to steal the elixir, Chang'e drinks it and flees back to her home in the heavens, choosing the Moon so she will still be close to her husband on Earth—who continues to revere her. (In a darker variation, her husband, the archer, after his heroic feat, becomes a tyrant. When she drinks the elixir and flees to the Moon, he attempts to shoot her down with his arrows as he once had the nine suns. He fails, but a goddess punishes Chang'e, turning her into a three-legged, albeit immortal, toad with a jade rabbit as a companion in her moon palace.)

While Elon Musk once planned to land a terrarium on Mars, the China Space Administration's Chang'e 4 unmanned lunar lander, which touched down on the Moon's far side on January 2, 2019, included a small greenhouse primed with seeds for cotton, potatoes, and rapeseed. That same month the Chinese government broadcast images of the first seeds sprouting, making it "the completion of humankind's first biological experiment on the moon." Another novel species also has landed on the lunar surface. On April 11, 2019, the Israeli Aerospace Industry's unmanned lunar lander, Beresheet—named after the first words in the Hebrew bible "in the beginning"—crash landed. Its cargo included the lunar library of the Arch Mission Foundation—and yes, tardigrades.

Inspired by Isaac Asimov's *Foundation* novel series, in which an interstellar group guards knowledge through a long dark age, the Arch Mission Foundation's stated goal was to create "Humanity's Backup Plan." More specifically, they labeled their "billion year archive" as "the first practical initiative with potential to guarantee that our species and civilization will never be lost." They plan to send out many libraries. (In addition to the one that crashed on the Moon in 2019, another was included in the glove compartment of the Tesla Roadster that SpaceX launched to Mars in 2018.)

The Foundation library has encoded not only cultural artifacts and data, like those that the Voyager spacecraft continue to carry into interstellar space, but also biological artifacts, including approximately 100 million cells from twenty-five humans and other organisms, encapsulated in artificial amber. These were mixed with microscopic fragments and relics from sacred places. The library also included, etched onto thin sheets of nickel, thirty million pages of text (including Asimov's *Foundation* series), as well as thousands of living (though dehydrated) tardigrades—the hearty creatures about the length of a grain of salt shown to survive in the vacuum of outer space and at temperature extremes.

As the fiftieth anniversary of the Apollo 11 mission approached in 2019, news of the Beresheet's crash and its cargo gained digital momentum. *Tardigrades had been smeared on the Moon.* The Foundation reported that many of the animals may have survived, although they would need water to be restored to life. They could be regarded as the Moon's first true colonists. The Twitter account of Arch Mission's CEO Nova Spivack spewed musings with hashtags such as:

#BringThemHome #TardigradeRescue, and #tardigradehaiku. Some examples:

Does anyone want to fund a Tardigrade rescue mission?

As well as:

Many tardigrades won't speak to me now. But the ones who still do, tell me that this is the best coverage their species has had in 500 million years.

Another tweet:

Schrodinger's Tardigrades: Earth's future is bifurcated into two macro-scale quantum states: one where the Tardigrades on the Moon are alive, and one where they are dead. Presently they are in an indeterminate state and therefore so is the future.

And the haiku:

Moon bears lost at Sea/Of Serenity; just dust!/Genesis boot disc.

Yes, a woman—perhaps from the China National Space Adminis-tration?—stepping on the Moon in the next decade will undoubtedly promote a new Moon fervor, but for now why not give the legion of tardigrades, that tiny hibernating Space Force, out on the Sea of Serenity, its moment of glory?

# CHAPTER NINE

# GOING INTERSTELLAR

• • •

"The search for knowledge, said a modern Chinese philosopher,
is a form of play. Very well: we want to play with spaceships."
—Arthur C. Clarke, "The Challenge of the Spaceship" (1946)

"Everything was beautiful, and nothing hurt."
—Stoney Stevenson, after being launched into the Chronosynclastic
Infundibulum, "*Between Time and Timbuktu*" (1972)

Although Robert Goddard, with his bald head, light moustache, and somber attire projected the image of a sober-minded scientist who happened to be interested in rocketry, but *not* in interplanetary shenanigans à la Edgar Rice Burroughs ("No one said the rocket would go to the planets, I simply want to make meteorological observations!"), he had other motives when, in 1918, he sealed an envelope, inscribed "Special Formulae for Silvering Mirrors." Before traveling west to conduct World War One weaponry research, he stored the envelope in a friend's safe; inside it was his precis for the emergence of interstellar humanity, "*The Ultimate Migration. The notes should be read thoroughly only by an optimist!*" The notes show that far from staid in imagination, Goddard had never given up on his youthful dream of blasting off from a cherry tree to outer space—he was an American cosmist, with an imagination as wild as Tsiolkovsky and company, and several leaps ahead of the golden age science fiction writers-to-come.

Goddard believed humanity needed to plan for an exit to the stars before the Sun cooled. He thought it might be wise to hollow out an asteroid or small moon to serve as the, preferably, atomic-energy propelled spaceship. He kept the details vague, but imagined many such spaceships sent off from Earth to various star clusters with the hopes that one or more would survive. The journey might take thousands or tens of thousands of years. Hundreds of generations might live on the ship during the journey, or the crew could be more sedate—he speculated it might be "possible to reduce the protoplasm in the human body to the granular state" to withstand the "intense cold" of interstellar space. In this case, the pilots would be awakened at, perhaps, 10,000 year intervals to adjust the course of the ship. The goal would be to reach a planet orbiting a viable star. If the technology for reviving desiccated humans wasn't effective, perhaps the vehicle would contain "protoplasm . . . of such a nature as to produce human beings eventually, by evolution." Tools, appliances, and descriptions of processes would accompany the protoplasm or the dried humans, "so that a new civilization could begin where the old ended."[1]

Unlike the potential ridicule that Robert Goddard faced for such ideas in 1918, when talk of spaceflight of any sort was regarded as fringe, scientists with interest in interstellar flight today do not have to mask their interest behind a bland "formulae for silvering mirrors." While NASA and private entrepreneurs are largely focused on projects within our solar system, interstellar fervor has been building, particularly since the 2009 launch of the Kepler Space Telescope. Before the Kepler launch, very few exoplanets, or planets orbiting other stars, had been identified. As of early 2020, courtesy of Kepler, about 4,100 exoplanets had been confirmed. By extrapolation there are likely to be as many as a trillion in our galaxy, among which, surely, could be a "Goldilocks planet"—not too close and not too far from its star so that water can exist in liquid form, a prerequisite for life, at least as known on Earth.

In recent decades, the general idea of preparing for interstellar travel has gained establishment credibility, particularly with the U.S. Defense Department's DARPA (Defense Advanced Research Projects Agency), which along with NASA was established in 1958, in response to the Sputnik launch. In 2011, DARPA hosted a "One Hundred Year

Starship" conference in Orlando, Florida. The goals were aspirational: to encourage invited technologists to a brainstorming session, assuming that this could trigger innovation, even if it did not produce a starship by the year 2111. The attendees, many of them with dual citizenship as scientists and science fiction writers, took at face value the proposal of readying a starship within a century. One participant, Louis Friedman, former JPL Advanced Programs leader and co-founder of the Planetary Society, recalled, "I asked the organizers to bring more focus to robotic flights" rather than "humans traveling to the stars." The result? Friedman noted, "I was outvoted 25–1."[2]

Why this urgency in 2011, regarding Goddard's "Ultimate Migration," one or two billion years before our Sun may fail us? Theorists of migration on Earth, such as those that brought "huddled masses" to the United States or to social media sites, insist there must be both "push" and "pull" factors. Applying these factors to outer space migrations, for "push" we have Earth's mounting environmental problems and the possibility of a cataclysmic asteroid or comet strike, disease pandemics, or nuclear warfare. For "pull" we have the thirst for adventure, curiosity, the profit motive, the recent discovery of many potentially habitable exoplanets, and the faith that venturing into the heavens fulfills cosmic destiny.

And so, (with apologies to Chairman Mao), DARPA encouraged a hundred starships to blossom. The conference demographics reflected the roots of the idea of interstellar flight, which has always been a joint venture of scientists and science fiction writers, with the general template for the starship developing along with the "space opera" in the 1920s. (Named after "horse operas," or westerns, in space operas, heroes such as Buck Rogers, Space Cadet Tom Corbett, and Han Solo maneuver dashing spacecraft that rely on mysterious "Super Science" propulsion systems, as they battle bad guys and monsters.)

The starship took shape on the covers and inside pages of magazines such as *Amazing Stories, Science Wonder Stories, Planet Stories* ("Strange Adventures on Other Worlds"), and *Astounding Science Fiction*. A starship might be: (1) for reasons pseudoscientific, faster than light, and useful for Anglo-Saxon heroes rescuing futuristic damsels from intergalactic Bug-Eyed-Monsters (BEM); (2) an intergenerational "world ship" (perhaps so large that it contains intact ecosystems) on which hundreds of generations could be born, live full lives, and die en route to the stars; (3) a more sterile spaceship of the ocean liner sort with crews uniformed in shiny jumpsuits, or, alternately, placed in suspended animation and periodically awakened; (4) a "sentient" starship, with AI or human/mech generated awareness; or (5) the related "bioship," an organic spaceship, with hull and other aspects coaxed from the genetic material of exotic species. Further, the entire ship might be a living organism—transporting others or simply a lone agent imbued with consciousness.

With the help of movies such as *Silent Running* (1972), the notion of starships as giant habitats, enclosing cities, rainforests, and farm fields emerged. In many science fiction novels, such as Brian Aldiss's *Nonstop* (1957) and Gene Wolfe's trilogy *The Book of the New Sun* (1980–81), the inhabitants of world ships, full of decaying cities and landscapes, and themselves somewhat decadent, have no idea they are traveling through space. Kim Stanley Robinson's *Aurora* (2015) was set on a starship with twenty-four linked habitats, a multicultural cast, and an AI guidance system. Its model of a self-sustaining, sentient, eco-starship relied in part on the concepts behind Gerard O'Neill's space colonies, and perhaps the Synergia commune's Biosphere 2.

As for current plans, some space enthusiasts doubt that human passengers will ever survive outside the solar system. Humanity may establish outposts on Mars and moons of other planets and asteroids but what lies beyond, these less-than-faithful suspect, is for virtual exploration. Accordingly, unmanned probes and posthuman crews have been proposed, as well as "panspermia projects" to seed comets, asteroids, and planets with basic life stuff. But for those who reject all limits on human endeavor, escaping the solar system and getting to the stars is not enough—the entire universe must be redeemed, its "dead stuff" transmuted into life and mind, a process that may expand to all the universe's atoms, so fulfilling a cosmic mandate (handed down from generation to generation via mad scientist speeches in B movies).

One of the great heroes of star settlers is John Desmond Bernal, a Cambridge University educated physicist and molecular biologist, who, in 1929, at age twenty-nine published *The World, the Flesh, and the Devil: An Enquiry into the Future of the Three Enemies of the Rational Soul*. Despite the racy title, this was not a standard bodice ripper, but rather the scientific version of one, depicting humanity's ultimate conquest of a yielding Nature. Bernal's specialty was X-ray crystallography, and it seems fitting that he remains one of the most influential of crystal gazers, that is, soothsayers—even Arthur C. Clarke conceded that Bernal beat him to many ideas.

In Bernal's lexicon, "the World" refers to inert matter and brute Nature and the challenge they pose to humanity's continuing conquest or wresting of knowledge and power therefrom. "The Flesh" refers to the limits of the human body and how it must eventually be augmented and reshaped into more efficacious and all-but eternal forms. "The Devil" of the title includes the emotional, psychological, and social forces that hold back human creativity, thought, and progress. A Marxist and, presumably, atheist, Bernal foresaw humanity becoming godlike, gaining pure dominion over nature. He had great confidence in human rationality, unlike Mary Shelley or the makers of 1950s schlock horror films in which knowledge-besotted scientists unleash giant mutant ants, bees, and blobs on the world.

In *The World, the Flesh, and the Devil*, while Bernal leaves the specifics of rocketry design to future discoveries, he does describe how a "space vessel spreading its large, metallic wings, acres in extent, to the full, might be blown to the limit of Neptune's orbit." To increase its range, such a solar sail craft might orbit close to the Sun for a slingshot effect. Wishing to avoid gravitational wells, Bernal rejects the idea of colonizing planets, proposing instead that humanity's outward expansion rely on orbiting colonies, later termed Bernal spheres (precursors to O'Neill's space colonies). These spheres, perhaps ten miles in diameter, would be built from material drawn from asteroids, planetary rings, or other detritus. Like Goddard, Bernal suggests that an asteroid might be hollowed out to serve as an initial shell. The globe's inhabitants would include workers, guests, and scientists. Not everyone may be suited to this way of life. People who prefer the joys of "primitive nature" to discourse and discovery could always remain on Earth.[3]

According to Bernal, sooner or later, the failure of the Sun will prompt some of these many colonies to veer off from the solar system. He is optimistic that technologies would eventually be developed to enable this. "Once acclimatized to space living, it is unlikely that man will stop until he has roamed over and colonized most of the sidereal universe." He justifies this by proposing that humanity, which is currently subservient and dependent on nature, must give up its "parasitic" role; oddly, his ensuing description sounds highly parasitic: "Man will not ultimately be content to be parasitic on the stars but will invade them and organize them for his own purposes." The stars, he believes, can be reconfigured, noting, "By intelligent organization the life of the universe could probably be prolonged to many millions of millions times what it would be without organization."[4]

That is not all that Bernal would reorganize. Our watery, soft bodies are just not that well-designed. Bernal comments, "In the alteration of himself man has a great deal further to go than in the alteration of his organic environment."[5] He imagines "radical alteration of the body" or even humans entirely freed from cell-based bodies—or "mechanized humans" with heightened sensory apparatus that might survive three hundred to one thousand years. While he admits the first prototypes may not improve on the life of a current garden variety human, "he would still be better off than a dead man." Even the brain might, cell be cell, be replaced by a "synthetic apparatus" that "would not destroy the continuity of consciousness" and ensure "a practical eternity of existence." He suspects that in this future, ultimately, "the direct line of mankind . . . would dwindle . . . perhaps preserved as some curious relic."[6]

Bernal calmly set out the ultimate big science agenda: humanity could conquer and transform the entire universe. He made the pursuit of progress, that is, the conquest and manipulation of all nature to aid humanity, into extreme sport. Not only would the human migration go on as scheduled, we would achieve immortality as a species and near immortality as individuals. Bernal's heady futurist cocktail has had enduring appeal, not only to space enthusiasts, but also to current Silicon Valley billionaires who embrace not only spacefaring but Fedorov's dream of conquering death. Among hopeful immortalists as of this writing are billionaires sponsoring bio-genetic research on longevity such as Larry Ellison, founder of Oracle; Sergey Brin, Google co-founder; Peter Thiel of Paypal; Ray Kurzweil, chief engineer at Google; and Bill Maris, CEO of Google Ventures. The future, according to this viewpoint, should be regarded as "limitless." If your goal is to transcend limits, death is an affront. Engineers can always find a work-around. Death just another program to be hacked. The most gung-ho of space settlers, like Fedorov before them, appear in full death denial—or revolt.

• • •

There are, not surprisingly, a lot of uncertainties in turning interstellar voyaging of any sort into a reality. As Philip Lubin, physicist at the University of California, Santa Barbara has noted, "The problem with a lot of the plans for interstellar travel is that they depend on key elements that are almost entirely speculative. You have to say 'insert miracle here' to get there." The biggest issue beyond propulsion, presumably, is life support, which is why many space boosters champion robotic missions. Yet most enthusiasts, like Bernal in his day, assume that if the human species and civilization stick around long enough, scientists and technologists will resolve the details. This is good enough for followers of Tsiolkovsky's dictum that "The Earth is the cradle of humanity, but mankind cannot stay in the cradle forever."

Post-Apollo, while NASA limited human missions to near-Earth orbit, some scientists, thinking big, continued to dream of interstellar travel. Rather like the cathedral builders of medieval Europe, they were willing to begin planning and research for projects that might not take genuine shape for centuries. No one expects interstellar spaceships bearing human passengers will be launching any time soon. Nevertheless, in 1989 NASA consultant and engineer Eugene Mallove and physicist Gregory Matloff published *The Starflight Handbook: A Pioneer's Guide to Interstellar Travel*. With its mix of technical explanations and advice to interstellar travelers, it suggests the sort of text James T. Kirk, Benjamin Sisko, and fellow space cadets might be handed at the Star Fleet Academy.

Dedicated to Konstantin Tsiolkovsky and Robert Goddard, the *Starflight Handbook* is a product of the 1980s, when *Star Wars* movies first were exciting audiences. The opening epigraph is a Yoda-like verse from eighteenth-century poet Edward Young, "Too low they build, who build beneath the stars." The introduction includes ultimate migration exhortations such as, "Ours should become a civilization that can outlive its star . . . It is never too soon to start thinking about interstellar expansion for the long-term preservation of terrestrial life. The time is now!"[7] Their handbook also makes clear how provisional was all such planning: most of the technologies required for the creation and successful launch of a generational starship do not—and may never—exist.

The *Starflight Handbook* offers different scenarios for expansion into the solar system and beyond, based on the future feasibility of speculative technologies and the prevailing political context on Earth, for example: "Scenario 1. Nuclear disarmament within 25 to 50 years; Earthlike worlds (ELWs) are discovered . . . Scenario 3. Slow human expansion into the Solar System . . . Discovery of many ELWs. . . . Scenario 6. An independent human civilization develops in the Sun's comet belt: ELWs are not found."[8]

For those ready to pilot a starship, the handbook also covers the mechanics of interstellar navigation. Correcting for relativistic stellar aberration—that is charting the position of one's rapidly moving vehicle in relation to distant and seemingly unmoored stars—will be one challenge, although "orders of magnitude more tractable than starship propulsion."[9] (How would this relativistic aberration be experienced if you were on the bridge of the *Enterprise*? Nikos Prantzos has explained that as a starship approaches light speed, stars brighten, while at 90 percent light speed (0.9 c) the field of vision shrinks—emptying out on the sides and sliding forward—intensifying directly in front. Navigating visually, clearly, will be impossible.)[10] A scientist friend of mine explained such relativistic effects, "Nothing with mass can get exactly to the speed of light. Relativistic speed indicates a significant fraction of the speed of light where one would begin to see the bending of spacetime due to the high speed. This is called a Lorentz Boost—when you shift to a sufficiently high velocity frame where your spacetime axes actually deform."[11]

Mallove and Matloff also evaluate propulsion systems necessary for interstellar travel from the likely to the highly speculative. Chemical propellants such as liquid oxygen are fine for jaunts to the Moon and back or even a one-way trip to Mars, but cannot get one much farther without refueling. Physicist James Benford has pointed out that a chemical rocket would not even be able to break free of the gravity of a "super-earth"—in this case a planet with a radius 50 percent larger than ours.[12] To fuel a chemical rocket to Alpha Centauri would require more atoms than available in the universe.[13] In fact, relying on chemical energy, all the mass in the universe could not accelerate one proton to relativistic speed. Interstellar propulsion theorists would be delighted to achieve even 10–25 percent light speed (0.1–0.25 c).

What are alternatives to fuel-burning rocket engines? In electric ion propulsion, electrons are stripped from elements such as mercury or xenon gas in plasma form and then accelerated through an electromagnetic field. Although the initial thrust is minute, with steady acceleration and a specific impulse ten times that of liquid chemical propulsion (giving it, effectively, a much higher "push per gallon" efficiency), craft with ion propulsion can gain greater speeds. Ion propulsion, which Robert Goddard began testing in the 1920s, is already a reality—some mini-satellites rely on ion thrusters to position themselves, and NASA's Gateway plans called for vessels to leave the Moon's orbit for Mars via ion propulsion. But ion engines, though capable of reaching higher velocity than chemical-propelled rockets, and useful for interplanetary ventures, cannot achieve anything near relativistic speeds and are not practical for interstellar travel.

It had long been an assumption in rocketeering circles that nuclear propulsion would improve on the woeful efficiency of chemical propulsion (Goddard speculated about this in 1918). From 1958 to 1965 a group of scientists that included Ted Taylor and Freeman Dyson worked on a prototype fission-powered rocket code-named Project Orion. This "nuclear pulse" rocket relied on exploding thousands of conventional nuclear weapons for propulsion; testing suggested that Orion was, surprisingly, safe. An "ablative plate" above the explosions would protect the rocket and passengers. Orion was designed with a trip to Mars in mind. But while Orion could handle interplanetary jaunts, it would require a supply of plutonium beyond its carrying capacity for centuries-long journeys to stars. The project was eventually killed over concerns due to fallout, dangers from crashes, and the mandate of the 1963 Partial Test Ban Treaty which forbade testing of nuclear devices in the atmosphere or outer space.

Of Project Orion, Dyson remarked to me, "It made sense fifty years ago. We had no idea how you could communicate in space. We thought of going out on Orion like Darwin on the *Beagle*. We'd step out on Mars with pads of paper and take notes. Five years later we would tell the world what we found. Now we have wide bandwidth communication and instruments on Mars supplying info. The whole big expedition approach became irrelevant."[14] This didn't deter Dyson from imagining, in 1968, an enormous starship that would carry 300,000 one-ton hydrogen bombs exploded every three seconds to accelerate to 0.033 light speed in ten days. It could make a trip to Alpha Centauri in 133 years. Including propellant, the massive starship would weigh 1.5 billion tons.[15]

A less explosive design, the nuclear thermal rocket, relies on a fission reactor core to heat a chemical such as hydrogen that would then create thrust through an exhaust nozzle. This engine also could not reach relativistic speeds, and so, would not be practical for interstellar flight. Another, more promising approach would use nuclear fusion.

In 1978, the British Interplanetary Society published plans for a working fusion power starship, Daedalus, a probe large as an aircraft carrier, which would have to be assembled in space. For propulsion Daedalus would rely on laser beams triggering fusion in solid pellets (of lithium hydride) containing deuterium and tritium. The ship, in theory, would be capable of achieving 0.12 light speed and make the 5.9 light-years journey to Barnard's Star in fifty years. Feasible? Not so far. Experimentation with the inertial confinement fusion planned for Daedalus began in the 1970s, and has yet to produce useful efficiencies. James Benford, one of the Starship Century participants, acknowledged that, as a technology, fusion is elusive. "So far fusion hasn't been conquered. Fusion has conquered us for sixty years."[16]

The *Handbook's* authors go on to consider Robert Bussard's 1960 proposal of the even more exotic ramjet fusion starship—which would rely on giant electromagnetic scoops to gather up protons from interstellar space to feed its fusion reactors, just as a whale might gather krill. Other wackier ideas: trapping small black holes to aid propulsion, antigravity machines (à la H. G. Wells), and the use of suspended animation for centuries-long flights. They also evaluate a "quantum starship" that relies on "ingesting" quantum fluctuations and converting this energy to propulsion. They emphasize that "out of this maelstrom of 'crazy ideas' may come the one we could use to get to the stars—in style!"[17]

Michio Kaku's *The Future of Humanity*, appearing three decades after the *Starflight Handbook,* also brims with optimism. Kaku, a physicist and science popularizer, leads with the good news. There are plenty of ELWs (Earth-like worlds). The galaxy, Kaku reminds us, is "teeming with habitable planets." The implied Q.E.D.: therefore human destiny is to be multiplanetary—or even interstellar. While Kaku adds caveats that some technologies are speculative, he has a solid hunch that interstellar travel, mind uploads, and the transhuman version of immortality, as proposed by Bernal, are in store. World, Flesh, and Devil all will be overcome. At his most bold, Kaku announces "Our destiny is to become the gods we once feared and worshipped" to which he adds, sounding a bit like the narrator of a 1950s' propaganda film, "The question is whether we'll have the wisdom of Solomon to accompany this vast celestial power."[18]

Kaku bases his optimism on an emerging "fourth wave" of technological innovation, loosely adapting his "wave" terminology from Alvin Toffler's *Future Shock* (1970). In the fourth wave (following steam, electricity, and computing), artificial intelligence, nanotechnology, biotechnology, new materials, and new propulsion technologies will coalesce to create an all but unimaginable future. Kaku enumerates all the potential propulsion systems for reaching relativistic speed, including fusion power, and then adds his own favorite. Recognizing that the "flesh" has its limits, he proposes, instead, "laser porting"—a technological version of Flammarion's "dream journeys" to other planets.

Kaku's laser porting depends on the possibility of uploading, or digitizing minds. He believes this will be the inevitable outcome of the "Human Connectome Project," a massive effort of the National Institutes of Health, begun in 2009, to map the human brain's entire neural networks. Like many of Silicon Valley's AI enthusiasts, Kaku believes that once the connectome is mapped, this will open up technology for uploading a digital recreation of a specific consciousness to a mainframe. He asserts that the uploaded connectome "in principle will contain all our memories, sensations, feelings, even our personality."[19] So step one, achieve immortality. For step two in Kaku's scenario, an uploaded personality, or consciousness reduced to an information stream, will be projected on lasers and sent at the speed of light around the universe. Your neural clone—presumably "you"—can be loaded onto distant mainframes in space, be assigned a mech body and continue exploring and sensing otherwise impossible worlds.

And beyond the speed of light is . . . faster-than-light (FTL) travel. One approach Kaku details would exploit wormholes. An advanced civilization might potentially create and maintain these still quite theoretical topological structures as shortcuts through space-time, allowing what amounts to FTL travel to distant galaxies, and possibly even journeys backward in time. He also considers the Alcubierre Drive, a physicist's take on the "warp drive" of science fiction. Miguel Alcubierre came up with this idea in 1994 while watching *Star Trek*. In Alcubierre's words, "By a purely local expansion of space-time behind the spaceship and an opposite contraction in front of it, motion faster than the speed of light as seen by observers outside the disturbed region is possible."[20] Calculations suggest it would require converting a mass at least as large as the planet Jupiter into energy to establish such a bubble. Some claim the energy of an entire star would be needed. Kaku comments drily that "it is still controversial whether such a starship can be built."[21]

. . .

Currently leading the laser-propulsion "starchip" effort at the University of California, Philip Lubin, commented, "I wish I had a warp drive, it's cool . . . But most of these proposals are not based on existing physics. The two most promising for interstellar propulsion are photonics, photon-based propulsion, and matter-antimatter. Though it hasn't been developed yet, the antimatter science isn't impossible." He added, "Our stuff [photonics] is really hard, but antimatter is even harder."[22]

Antimatter drives would require bringing protons and antiprotons into contact, with the ensuing reaction turning 100 percent of the interacting particles into energy. But not all that energy can be harnessed. One approach is to herd the charged subatomic pions released in the reaction through a magnetic nozzle for thrust. In other approaches, the explosion's high energy by-products might be exploited to prompt fission or fusion reactions. A major drawback to any such scheme: antimatter is scarce. Harvesting antiprotons from all the Earth's high energy particle accelerators would yield only about a trillionth of a gram yearly. Another drawback, by current methods, it requires more energy to isolate antimatter than it can provide.[23] Lubin commented that we not only cannot make enough antimatter currently but we cannot store it for much more than fifteen minutes. Lubin thought this may be a good thing since the most likely use, given humanity's track record, would be to use antimatter to build massively powerful bombs.

Most everyone interested in interstellar flight agrees that the most promising current approach is neither fusion nor antimatter but beamed sailing ships. The general idea of light sailing has a long pedigree. After observing the movement of comet tails and attributing it to a solar wind, mathematician and astronomer Johannes Kepler suggested in a letter to Galileo in 1608, "Provide ships or sails adapted to the heavenly breezes, and there will be some who will brave even that void." Kepler deduced, correctly, that light carries momentum and can transfer pressure to microthin sails.

In 2010, the Japan Aerospace Exploration Agency's Ikaros sailcraft was released from a rocket on a mission to Venus. It spread out its fourteen meter wide kite-shaped sail, which then was accelerated by photons from the Sun. The Ikaros sail was able to shift its flight path by electrically activating "liquid crystal panels" on its edge to adjust light pressure. More recently the Planetary Society's LightSail 2, launched in July 2019, was the first vehicle to use solar sailing to maintain Earth orbit. Sunlight could potentially propel sailcraft on a two-year flight to the Oort Cloud at the outer reaches of the solar system, but will not open the way to the stars. A greater boost of energy is needed.

Scientist and science fiction writer Robert Forward's *Flight of the Dragonfly* (1984) involved an enormous laser beam station near Mercury to propel a sailed starship to Barnard's Star on a one-way voyage. In 1985, Forward designed a microwave-propelled sailed star probe, Starwisp. Because microwaves have large wavelengths, the enormous sail could be a light metallic mesh. To propel Starwisp, a twenty gram probe, to relativistic speed (in this case 0.2 light speed), the microwave beam would have to be ten gigawatts, and the sail one kilometer in width. More problematic, Forward calculated that in space a 50,000 kilometer wide Fresnel lens (functioning like those on lighthouses, but in reverse) would have to be built to focus the microwave beams.[24]

James Benford explained that Forward put his hypothetical beaming station near the Sun to collect solar energy, which necessitated the Fresnel lens. He offered as an alternative, a beaming station on or near the Earth to prevent the need of the giant focusing lens.[25] But the costs and electrical demands made it unfeasible. In his 2013 paper, Benford calculated that a four-ton starship probe—with most of the mass in the sail—relying on microwave beaming could accelerate to 0.1 light speed. To do this would require nearly 500 terrawatts (500 trillion watts) of power ("such propulsion energies may not be available for centuries"), a ten-kilometer-wide sail, and cost a minimum of $42 trillion (about three times the Earth's current wealth generated yearly). Overheating of the sails during acceleration was a concern. It also was not clear how such a ship would stop once accelerated—Benford's examples were for "fly bys." His conclusion? "Missions will not be flown until the cost of electrical power in space is reduced orders of magnitude below current levels."[26]

Philip Lubin's 2015 paper "A Roadmap to Interstellar Flight" renewed interest in "beamer" technology. His roadmap proposed large arrays of lasers brought into phase to launch fleets of micro-sized "starchips." Such technology was not out of reach. While fusion research has moved slowly, and chemical rockets have changed little in efficiency since the development of the liquid chemical V-2 rocket during World War Two, Lubin noted that photonics has undergone exponential growth, doubling every 1.7 years in efficiency, a progression similar to that in computing power expressed in Moore's Law.[27]

In 2016, Lubin's "Roadmap to Interstellar Flight" caught the eye of Pete Worden, executive director of Breakthrough Initiatives. According to the Breakthrough website, the organization, funded by billionaire entrepreneur Yuri Milner—himself named after famed Soviet cosmonaut Yuri Gagarin—is "probing the big questions of life in the Universe: Are we alone? Are there habitable worlds in our galactic neighborhood? Can we make the great leap to the stars? And can we think and act *together*—as one world in the cosmos?"

Worden concluded that Lubin's approach was more likely to "make the great leap to the stars" than antimatter or fusion. Endorsed by Stephen Hawking, and with total funding of $100 million, the organization launched Breakthrough Starshot. The goal, as in Lubin's "Roadmap," by 2040, to launch minute (one-gram) "nanocraft" or "starchips" with one-meter lightsails at 0.1 to 0.2 light speed to the Alpha Centauri triple-star system. These tiny starships, packed with sensors, if all went as planned, would arrive at Alpha Centauri roughly within twenty years of launch and send back data.

Currently with grants from NASA, Lubin and his team were hoping to better the Starshot goal and achieve 0.25 to 0.26 light speed with their starships. Pushing open a nondescript door on the seaside campus, I entered their test lab, cluttered with electronic equipment, where postdoctoral fellow Prashant Srinivasan was at work. Wearing a plaid shirt and glasses Prashant asked, rhetorically, "Why not use a giant laser?" He paused, "To propel even tiny objects to Alpha Centauri needs gigawatts of power. We can't find a gigawatt laser. They have not been created yet. Our approach is to leverage existing technology and combine light from small sources, coherently. Therein lies the rub."

Obstacles include diffraction that disperses light as well as other sources of "noise" that might prevent the numerous laser beams from converging at the target with high efficiency. The UCSB team's solution is to rely on a guiding or "seed laser" that is sent from the target back to the laser array. Srinivasan explained, "The reverse laser sends information about phase fluctuations, and we apply that information to bring our lasers into phase."

Gesturing to a small electronic laser arranged on the work table aimed through several lenses at a small solar cell on a stand, I asked him, "How many of these would you need to propel a one-gram craft to Alpha Centauri?"

Srinivasan answered, "One billion."

For basic research the UCSB group was testing off the shelf lasers as well as high-efficiency solar panels developed at JPL to receive the photons. Ultimately, they will scale up the technology. To propel starchips to Alpha Centauri will require an array as large as one square kilometer in the desert, which would beam as much as 50–70 gigawatts. While it would be preferable to have the array on the Moon or in orbit, Lubin explained that going off planet is "much more expensive to be practical now." The Earth-based array, once in place, could launch as many as 40,000 starchip probes a year. Although the Earth's atmosphere adds turbulence—remember all those early contrasting maps of the surface of Mars?—adaptive optics systems, like those Srinivasan described, can mitigate this problem.

If they work out the propulsion system and a viable nanocraft, the other main obstacle, Lubin noted, is "the interstellar medium itself. There are particles in it, dust, protons, that you have to plow through." At relativistic speeds even the impact of one proton or speck of dust could be disastrous. How devastating is dust to interstellar travel? Lubin said, "The simple answer is that it looks pretty hard but it doesn't seem fatal."[28]

Lubin added that they are not only working on wafer craft but could also scale up the project so that photonics could be used for human interplanetary travel, permitting, for example, an Earth to Mars trip in thirty days. In addition to laser propulsion of sails, lasers could beam power to ion engines on spacecraft, easing payloads. Such a craft could be slowed by reversing the ion engine direction. As of 2019, Lubin's team had a NASA grant to explore sending beams from the Earth to the Moon to power rovers or from the Moon to power ion engine craft en route to Mars.

Was he concerned that some less-than-noble actors might use orbiting or Earth-based lasers as weaponry? "Absolutely. Clearly you'd have to be asleep to think that someone might not think 'what can I do with eleven gigawatts of laser power other than nudge meteors?'"[29] Photonics might also create a new version of the Cold War arms race. That might be one argument, he suggested, for situating a laser array on the far side of the Moon, where it could not aim at targets on Earth.

Asked if humans will ever travel in interstellar space, Lubin was less enthusiastic than Mallove or Kaku. For starters, no one knows how to stop such a starship after it has accelerated to relativistic speed. "It's good for sci-fi. The idea of interstellar travel is very romantic and can be philosophical. And our tech could be scaled to send humans out. But there are lots of problems. My feeling is humans are high maintenance and a waste of time. Humans are too big for what they can do. They are frail. They can't be put into a deep sleep state. They are not suited for this stuff. It's like choosing a pet. It's better to have a cat than a whale in your house. We should be thinking about post-humanity. We can't make a Million Dollar Man yet but I think in one hundred years we could have a decent AI structure. The question becomes then, what do humans have to offer? Perhaps the answer is that we are creators and we send out our AI emissaries."[30]

• • •

Others in the field agree that "beamers" are the most promising near-term technology for interstellar travel. James Benford, an advisor to Milner's Breakthrough Starshot, has noted that in principle beamers can exceed speeds achieved via fusion engines. He added that, "The best feature of beaming energy is that the Beamer—with all its mass and complexity—is left behind, while the relatively simple ultra-light sail, carrying its payload, is driven far away." Once the cost of the infrastructure is covered, the same beaming station can launch an entire fleet of starships. "The most likely alternative, a $100 billion fusion rocket, is far more expensive because you throw the whole rocket away for each flight."[31]

Freeman Dyson was also in accord with the notion that "you can't take it with you." Propulsion sources should not be on board the craft. Dyson has proposed that, in about a thousand years, to propel craft, humans could establish a network of high energy beams through space, whether laser, microwave, or pellet—this latter involves a stream of fuel aimed at a craft that then is ionized and converted to thrust. While a heavy-duty system could transport humans, a light-freight system would carry packages of information.

He likened this system to a public highway. In a telephone interview, Dyson said, "I'm sure it will be feasible someday. The reason public highways are effective is because they are cheap. The cost is shared between huge numbers of travelers. Same in space. If everyone builds their own ship this is very costly. Only big organizations can go. It is not available to everybody. For the public highway a laser beam or microwave beam could provide propulsion. The energy would be from starlight or solar sources. This could work within the solar system or beyond."[32]

But what will life be like on a starship if such a vehicle can ever be launched? DARPA's "One Hundred Year Starship" project inspired the group Icarus Interstellar, founded in 2011, which until 2017 held semi-annual Star Ship Congresses. One Icarus Interstellar project, Project Persephone, led by Rachel Armstrong, a professor of architecture, was charged with developing a "habitable starship architecture."

Armstrong's book *Star Ark: A Living Self-Sustaining Spaceship* (2017) focuses on the cultural dimension of a spacefaring civilization. Her goal is to examine "the Interstellar Question—i.e. the idea that we may one day live beyond the world we know and settle distant planets . . . the book accordingly seeks to provide a form of cultural catalysis by which an interstellar culture may be seeded."[33] Her main argument is that creating a star-going culture is as critical as engineering. To support her point, she notes, "It is reported that being inside the ISS is analogous to living in a sweaty portable toilet."[34]

She presents herself as no fanatic. Miracle technology breakthroughs may or may not come. To her the Interstellar Question, whether humanity will ever colonize the stars, is not a given. She instead writes of "constructing alternative modes of existence," and focuses on the idea of creating a "living architecture" that doesn't rob an ecosystem of resources, but instead generates its own ecology, its own tissues that combine nature and technology into "supersmart living fabrics." She gives the example of "living walls" that replenish air, recycle water, and fertilize plants. Another of her notions was architecture that can grow and self-repair. (One application on Earth she proposes is to "grow" an artificial reef under Venice using "smart droplets" that can be chemically stimulated to slow that city's sinking.)

Armstrong wishes to provide clues on how humanity might transition from the Anthropocene, that is an age in which humans dominate the biome and have triggered climate change and mass extinctions on Earth, into an Ecocene, an age that is more in balance, and less dependent on the competition and resource-spoiling model of capitalism. She argues that reaching for the stars, an effort that potentially will fail, to be justified, must immediately help us take a "greener" approach to the surrounding environment, whether on Earth or in outer space. This idea is reflected in some of her gnomic sayings, such as, "we must consider the fabric of space not as being a sterile site or an industrial resource fit only for plundering, but as a living, potentially fertile system."[35]

The rare glimpse of specific starship plans in *Star Ark* skews to prankish conceptual art. Architect Teodor Petrov's proposal for a "space skyscraper" follows Armstrong's suggestion that "creating the conditions for livability as we travel to the stars may be achieved in a very immediate and practical sense by weaving starship fabrics into our cities."[36] Petrov's space skyscrapers would "grow" on site in cities by extracting materials from the air in the form of smoke, dust, and other particles. The building would form top down, beginning with its tower tops. It would maintain its own self-sustaining ecosystem, with "living walls" to clean and circulate water and air.

The space skyscraper would slowly take shape—Petrov's renderings suggest its final form as a tendriled, vinelike, metallic combination of Watts Towers and Early Flash Gordon. People would live and work in its laboratories "as if they already inhabited an interstellar craft." Petrov proposes growing these skyscrapers in Beijing, Mumbai, Cairo, Orlando, Mexico City, and Honolulu. When no longer useful on

Earth, relying on fusion propulsion, the building would be launched into outer space. Materials from these buildings then can be scavenged as scaffolding from which to build a starship.[37] Armstrong comments that Petrov's project is "largely speculative" but she likes the "principle of not tearing down perfectly good buildings into rubble—plus the idea of prototyping and alpha testing space structures for habitation on Earth before they are inhabited in an orbital context."[38]

Armstrong's vision for interstellar settlement also includes panspermia, that is, launching into space bacteria, microbes, algae, fungus, and plant life. The idea is to "seed" space with simple ecosystems, to make it, eventually, more habitable, or at the very least, aid it to sustain some forms of life. Armstrong turned here to an idea of Freeman Dyson's. The hypothetical Dyson Tree is a genetically modified plant species that might thrive inside a comet and eventually provide breathable air to humans or other biota. Dyson had also proposed that custom-made "eggs" equipped with the makings of a biome could be sent to outposts.

Dyson's approach assumes that we cannot count on finding "Goldilocks" planets, nor should we want to. As he said to me, "Most of the real estate in the universe is small, cold. The media always talks about planets, but smaller objects which you can easily land on and take off from are much more convenient. They are not Goldilocks at all. Very cold and dismal at the moment. You have to change the environment by building comfortable habitats for people and life in general."[39]

And about those exoplanets—any notion of planetary protection from biological contamination, as in current treaties, would have to be evaluated. Barren worlds, unquestionably, are fair game in Armstrong's mind. She cites scientist Michael Mautner, a chemist and astrochemist, who has argued that such panspermia projects fulfill a human mandate to "enliven our cosmos." Mautner holds the view that life on Earth might originally have been seeded by meteor or comet strikes, and that we should return the favor. To "contaminate" might instead be to seed, enrich. Even if we never succeed in venturing to the stars, other life can develop out there, with our help.

• • •

Not all space enthusiasts think the push for a human presence throughout the galaxy is within our capabilities, no matter how appealing on the pages of science fiction novels or movie screens. Astronomer Martin Rees has observed that "nowhere in our solar system offers an environment as clement as the Antarctic or the top of Everest. Space doesn't offer an escape from Earth's problems."[40] While gung-ho about space exploration, Louis Friedman sees no hope for sending humans beyond the solar system. Regarding starship propulsion, he says the concepts have not changed much for the past fifty years, and the most promising remain "not yet invented." Likewise, while it is fun to describe placing crews in suspended animation and then "magically" reviving them centuries later, this again requires a possible miracle. Friedman pushes the unpopular idea that we may have to accept some limits on human achievement. Starships with human crews, whether graduates of Star Fleet Academy or some other diploma mill, will always remain pure fiction.

As Friedman puts it, for humans "getting beyond Mars is impossible— not just physically for the foreseeable future but also culturally forever."[41] Home for us might extend from Earth to Mars and, perhaps, the inner planets. He remarks that even this is iffy, and recalls all the *Popular Science* articles in the 1950s about undersea colonies housing hundreds of thousands of people. The engineering of such colonies would be an undertaking far less costly than colonizing space, but has never been in the works. As for space exploration and travel, robots do it better. According to Friedman, the space race was never between the United States, the Soviet Union, or other nations, but between humans and the robots. The robots have won. He argues that the "long-range future of humankind is to extend its presence in the universe virtually—with robotic emissaries, bio-engineered payloads, and artificial intelligence."[42]

Other true believers willing to acknowledge that human bodies are not engineered for long space journeys often propose reshaping humanity into transhuman forms better suited to outer space. But they also bring in one more possible strategy. They wonder if human DNA could be replicated inside hardy bacteria which then could be whisked to distant worlds on a starchip or perhaps Dyson's highway. At its destination, the DNA code could be "printed out," creating an instant human or, at least, embryo. One such enthusiast was JPL's Curiosity Rover team leader, Adam Seltzner, who announced at a 2014 conference that "Our best bet for space exploration could be printing humans, organically, on another planet."[43]

And while it may be possible, one day, to transport or transmit the human genome into deep space, there is no way now, nor may there ever be, to print humans. It is good in science fiction though. Greg Bear's *Hull Zero Three* (2010) is a tale of a generational ark that includes narrative sections titled "The World," "The Flesh," and "The Devil." This starship, built around an icy moon, and equipped with a "boson drive" goes amok as it approaches a promising though inhabited exoplanet. Warring factions on the ship and aliens who really don't want intruders on their turf hack the ship's genetic printing of cloned humans and monstrous beings (programmed for wiping out aliens); the starship crew includes, as an Adam-like figure, an unthawed "Teacher" trying to make sense of it all.

• • •

Human interstellar travel remains, as Goddard once pointed out, "for optimists." Nurtured in science fiction, the idea just may remain limited to science fiction. This can be a disturbing thought to the science fiction faithful, as it has been a commonplace that its writers are prophets. Hugo Gernsback, who founded *Amazing Stories*, the first science fiction magazine, in 1926, offered as its motto "Extravagant Fiction Today— Cold Fact Tomorrow." A more prosaic variation on the Gernsbackian dictum, attributed to Isaac Asimov: "Science fiction today is science fact tomorrow." (When I was teaching adult education in Queens, New York, Asimov's home turf, one of my older students, a rugged senior citizen who had likely weathered the prison system, liked to quote this, and when I questioned it as a universally valid statement, grew furious. His stare, rather like a Venusian death beam, actually hurt my heart. Clearly, I wasn't as faithful a science fiction fan as he.) Director Ridley Scott, falling in line, titled his eight-episode documentary of the genre *The Prophets of Science Fiction*.

And, of course, there is truth to the cliché. We need look no farther back than the 1946 advent of Dick Tracey's two-way wrist radio to recognize antecedents of current technology in early pop culture. However, not all that is conjectured is prophetic. And the space opera, alas, may be false prophecy. Kim Stanley Robinson, well-known for his 1990s trilogy (*Red Mars, Green Mars, Blue Mars*) in which he depicts the terraforming and settling of Mars, turned to interstellar spaceflight in his 2015 novel *Aurora*. Its twenty-sixth century world ship features twenty-four biomes, each so large that members of the clans living there rarely travel outside their realms. Following Dyson's ideas, Robinson's starship is not self-propelled, it is powered with an electromagnetic launcher—like an O'Neill mass accelerator—near Saturn's moon Titan, and then with laser beam boosts.

After seven generations traveling at the modest 0.1 light speed, the ship and its occupants begin to settle a moon of Tau Ceti, twelve light-years from Earth, with water and an oxygen-rich atmosphere. After initial forays to the new world, the human settlers die off, infected by prions (abnormally folded proteins that spread similar abnormalities in neural proteins) dispersed throughout the moon's ecosystem. The starship's survivors divide into factions: those who would go on to explore another potential planet, and those who would return to Earth's solar system.

Euan, a quarantined and doomed pioneer on the Tau Ceti moon, concludes that settling new worlds is a fool's errand, "they're either going to be alive or dead, right? If they've got water and orbit in the habitable zone, they'll be alive. Alive and poisonous. . . . So that's going to be a problem, most of the time anyway." As for finding a lifeless planet or moon to terraform, he argues, "Even if you've got the right bugs, even if you put machines to work, it would take thousands of years. So what's the point? . . . Why not be content with what you've got?" Trying to imagine the mindset of those who set them off on the journey, he added, "Who were they, who were so discontent? Who the fuck were they?"[44]

Euan goes on to cite Fermi's Paradox—that is, if millions or billions of planets are likely hosts of life, why haven't we heard from or detected other intelligences? He argues, "Fermi's paradox has its answer, which is this: by the time life gets smart enough to leave its planet, it's too smart to want to go. Because it knows it won't work. So it stays home. It enjoys its home."[45] When this starship does the unthinkable, and returns to Earth, its crew is met, largely, with hostility. The returnees scroll through the news and read how they are cowards and traitors who have "betrayed the human race, betrayed evolution, betrayed the universe itself."[46]

Just in case Robinson was playing the devil's advocate—perhaps to deflate starship genre clichés, while secretly maintaining faith in such a future, I followed up with him. He confirmed that it was a critique. "*Aurora* does fairly sum up my views, although it's also the case that it describes a single example of interstellar travel, so it is not comprehensive." He deemed interstellar travel "unsolvable . . . By unsolvable, I mean there is no future technology or human development that I believe in that can solve them." *Aurora*, according to Robinson, was "a response to what you might call the cultural meme, or simply the idea, that humanity is destined for the stars, with sometimes the corollary that if we don't become a star-faring species, we are in danger of extinction or will have been some kind of failure as a species."

Robinson flatly rejected the cosmist notion of destiny. Based on more recent understanding of physics, biology, and ecology, he asserted it was clear that star settling was impossible. He added, "The new paradigm might be that life is a planetary expression, and away from its home planet, life withers and dies." Robinson had no objection to space exploration, and in general liked NASA's slow, cautious approach of "stay safe, don't get people killed, use robots when you can. Use space for science."

He added, "The scientific possibility of starships is irrelevant when considering them as a story space. It would be like complaining that Middle Earth isn't real. So what? As a story space, they are perfectly legitimate . . . I am not trying to kill a science fiction subgenre that I have greatly enjoyed, but rather the idea that we could really do such a thing, and that we should, or we are failures. That I think is wrong."[47]

Others would like to kill the genre. When they issued *The Mundane Manifesto* in 2004, Geoff Ryman and a group of science fiction workshop participants thoroughly trashed the trope of the interstellar journey, along with time travel, teleportation, aliens, and alternative worlds. This manifesto (its name riffing on the science fiction community's slang term "mundane" to refer to outsiders) noted that "interstellar travel remains unlikely" and science fiction often "can lead to an illusion of a universe abundant with worlds as hospitable as this Earth" fostering a "wasteful attitude to the abundance that is here on Earth." The manifesto insisted that the most likely future was on this planet within this solar system, and, accordingly, it would be a relief for writers to focus "on what science tells us is likely rather than what is almost impossible such as warp drives. The relief will come from a sense of being honest." Not wanting to be utter killjoys, they added that other writers should "not let Mundanity cramp their style if they want to write like Edgar Rice Burroughs."[48]

If Friedman, Robinson, and the Mundanes are correct, while robotic space exploration is fine, it is time we try to focus on Earth-based problems rather than plan an escape to new worlds. There will be no "ultimate migration." The "pull" may be strong but we neither have the right bodies or the miraculous "push"—warp drives, et al., to make it happen.

# CHAPTER TEN

# THE METAPHYSICAL LURE OF DEEP SPACE

• • •

"They were savages . . . descendants of a research team of scientists that
had been lost and marooned in the asteroid belt two centuries before . . .
They called themselves the Scientific People."
—Alfred Bester, *The Stars My Destination* (1956)

"We are not just an insignificant species of semi-intelligent apes . . . we are the
sole source of consciousness in an otherwise dead cosmos. It is all up to us."
—Marshall T. Savage, *The Millennial Project* (1992)

To justify the spacefaring impulse, Manifest Destiny was once enough.
America would be settled from sea to shining sea—and then we could
add on upper stories. Nature and indigenous people were to be conquered, profitably, and remembered nostalgically. As 1950s television
broadcasts of *Gunsmoke, Death Valley Days, Maverick,* and *Have
Gun Will Travel* made clear, when Sputnik launched, the country had
been missing the wilderness. This syndrome had been diagnosed a
half-century earlier, in historian Frederick Jackson Turner's 1893 paper
"The Significance of the Frontier in American History." He announced
that the American frontier had closed, and with it, a source of virtue
and innovation. Turner argued that the frontier had nurtured American independence, creativity, and cooperation. Well into the twentieth
century, scholars could insist that the frontier generated self-reliance
and technological ingenuity, as they fretted about the shapeless, conformist society emerging.

Sputnik, then, was a gift. Outer space provided "a new frontier."
The same angel that decorated John Gast's 1872 painting "American
Progress"—a pre-Raphaelite maiden in windblown gown striding in

the sky before a westward march of hunters, Conestoga wagons, stagecoaches, plows, railroads, and telegraph cables, an amassed force scattering buffalo and bewildered Native Americans—now could float sanctimoniously below Saturn rockets, plummeting rocket stages, and heat-shielded capsules.

The New Frontier that John F. Kennedy spoke of in his 1960 nomination acceptance promised a renewal of past glory, and a rallying of national purpose in a time of "creeping mediocrity." After diagnosing "a change—a slippage—in our intellectual and moral strength," Kennedy reminded the audience of the fortitude of the pioneers and then proposed that a new generation had to step forward. "We stand today on the edge of a new frontier. The frontier of the 1960s, the frontier of unknown opportunities and perils, the frontier of unfilled hopes. . . . Beyond that frontier are uncharted areas of science and space. . . . I am asking each of you to be pioneers toward that new frontier."[1]

The speech reflected not only youthful resolve, but the link between outer space and the Old West then enlivening popular culture. In 1955, Walt Disney opened the southern California theme park Disneyland, segregated into four regions, with two of them "Frontierland" and "Tomorrowland" (complete with a Space Port). Likewise, Disney's television show had aired three "Man in Space" episodes from 1955 to 1957 that included rousing lectures on the future of spaceflight. As in Disney's realm, Kennedy's high-flown speech officially hitched the pulp western to pulp science fiction. The result: a politically-sanctioned marriage between the rugged individualism depicted in Marlboro cigarette advertisements and the galactic federations, interstellar trading posts, and warp-drive spaceships of science fiction magazines.

The idea of the pioneer spirit has since yellowed, become dog-eared, even moldy. Many scholars now insist the tale of western expansion was one of conquest, genocide, broken treaties, capitalism and exploited labor, slavery, suicide, depression, disease, ecological conquest, species extinction, and the pollution and plundering of resources. Yet this reframed narrative has seldom prevented U.S. politicians from evoking the pioneer spirit when promoting spaceflight, whether Lyndon B. Johnson, George H. W. Bush, George W. Bush, or, more recently, Mike Pence.

The general notion that hardship and challenge can bring out the best in individuals and communities has continued to tease at us. Early space advocates could insist that the New Frontier would (as it did) provoke technological innovations and offer us (as it has not) a grand destiny. By the time of the final Apollo missions, as many people might have chosen to view episodes of *The Beverly Hillbillies* as murky broadcasts of astronauts bobbing on the lunar surface.

Even post-Apollo, space advocates worked the frontier metaphor. In *The High Frontier* (1977), physicist Gerard O'Neill depicted plucky families one day setting off from a space colony on self-built craft or "wagon train" to homestead their very own asteroid. Off in the asteroid belt beyond Mars, he had them engage, metaphorically, in "clearing and stump-clearing," before celebrating with a Thanksgiving feast.[2] In *The Case for Mars* (1996), Robert Zubrin quoted Turner's frontier thesis to establish that western civilization "can only exist in a dynamic expanding state."[3] The frontier had vanished, but, like a missing limb, refused to go away.

More recent spacefaring proponents sidestep America-centrism —but not the concept of frontier dynamism. A society needed to expand—and rise—or fall. In *The Future of Humanity*, Michio Kaku referenced fifteenth-century China as a technological and cultural center that declined when a court faction halted the expeditions of Zheng He's great fleet of ships and largely ended foreign contact. (Zheng He's final expedition involved over two hundred ships carrying 27,000 men.) Kaku mused, "I sometimes think about how easy it is for a nation to slip into complacency and ruin after decades of basking in the sun . . . nations that turn their backs on science and technology eventually enter a downward spiral."[4]

With obvious signs of national decline in the United States when he wrote in 2018, including a growing disinterest in science funding, and a taste at the top for bizarre conspiracy theories, Kaku had a point. While government officials of the United States continued to question climate science and to loosen environmental standards, China was funding Big Science efforts including development of renewable energy, a 62-mile long super collider, artificial intelligence, and an ambitious space program. We could not, Kaku argued, afford to burn our ships and turn inward. That way led to ruin.

Spaceflight's popularizers, driven by a vision of a transformed future, have long been concerned with stagnation. As Bernal wrote in 1929, those not up to the challenge could stay behind on Earth. However, the environmental movement of the 1970s and beyond posed a challenge to the concept that embracing technology assured advance. The idea of progress itself became problematic. By the 1970s, not just the counterculture, but policy wonks were questioning the fairy tale model of endless capitalist expansion and its implied "happily ever after."

Grappling with such "anti-growth" sentiment of the 1970s, space boosters seized on a new trope: humanity did not need to learn how to make do with less and establish sustainable practices, it could simply move beyond the Earth's largely closed ecosystem. In his 1971 piece "The Extraterrestrial Imperative," for the *Bulletin of the Atomic Scientists*, German-born aerospace engineer Krafft A. Ehricke made the case for technological civilization's continued expansion without limits. Ehricke later described the Extraterrestrial Imperative as "a primordial imperative, bred into the very essence of the universe" that was "driving . . . the natural growth of terrestrial life beyond its planetary limits."[5] Humanity should extend its range beyond the Earth's "sensitive biosphere" and follow "the extraterrestrial imperative" into a limitless cosmos.

Like the cosmists Federov and Tsiolkovsky, and philosopher Henri Bergson who made statements such as "Man might be considered the reason for the existence of the entire organization of life on our planet,"[6] Ehricke infused evolution with purpose. We needed to take the next step and become spacefarers.

Evolution itself demanded humanity's expansion. Ehricke announced, "The Creator of our universe wanted human beings to become space travelers. We were given a Moon that was just far enough away to require the development of sophisticated space technologies, but close enough to allow us to be successful on our first concentrated attempt."[7]

Ehricke also presented the history of life on Earth as a series of willful responses to evolutionary crises. "The first great crisis," about 2.5 billion years ago, he argued, was the buildup of atmospheric oxygen toxic to anaerobic bacteria. Ehricke described the options available to Earth's early biome: "Give up and perish, regress to a minimal state of existence, or advance and grow."[8] Life's response? Species of aerobic bacteria emerged and flourished, leading, eventually to the Cambrian Explosion of animal phyla. Ehricke insisted that a "second great crisis" emerged in the late twentieth century when humanity's industrial technology unleashed toxic molecules in the ecosystem that could not be recycled. The answer was for "life" to respond by leaving the Earth's closed ecosystem, in this way offshoring dangerous industry to outer space.

Although clearly not an evolutionary biologist, Ehricke was not unusual in positing an underlying meaning to evolution. Even Charles Darwin had struggled with two portrayals of evolution: the first, stressing random variations, paints evolution as "a directionless process, going nowhere slowly," while the second version, steeped in Enlightenment thought and the Victorian ethos, assumes beneath that randomness an immanent, progressive force unfolding in history.[9] Darwin's apostles had no problem with the idea that, mechanics aside, the "purpose" of evolution was to produce "higher" species endowed with consciousness.

It was only in the mid-twentieth century, responding to the excesses of the eugenics movement and scientific racism that biologists sought to extract evolutionary theory from the notion of progress. Evolution must no longer prop up notions of racial or cultural hierarchies. Paleontologist Stephen J. Gould labeled the very notion of progress "noxious, culturally embedded, untestable." The age-old concept of a hierarchy of species, or Great Chain of Being also was noxious. Setting aside technical arguments, Gould commented, "Our geological confinement to a moment at the very end of recorded time must engender suspicions that we are a lucky accident, an afterthought rather than the goal of all creation."[10] Denying that we were a "chosen" species suggested there was no great plan to fulfill. Evolution might be a "directional" process but as a compass it was pointing nowhere.

True believers such as Ehricke, however, regarded the Extraterrestrial Imperative as a direct order, "bred into the very essence of the universe." The fact that we are destined for the stars, as it were, was written in the stars. Once one discerns "meaning" embedded in the fabric of the cosmos, differing interpretations will support alternate articles of faith. Such grand circular thinking merges science with myth. While Ehricke agreed it was humanity's choice to wither away on the Earth as its biosphere deteriorated or to take bold steps into the universe, he clearly presented the former as untenable and insisted outer space represented humanity's true destiny. In so doing, the Garden of Eden might be restored on Earth or better yet, we might turn the entire universe into our garden—recreating it to our specifications.

Linking spacefaring to evolution encouraged hope in unbounded human potential and the belief that humans were gods in the making. In a posthumous article, Ehricke enunciated "Three Laws," including: (1) "Nobody and nothing under the natural laws of this universe impose any limitations on man except man himself," (2) "as much of the universe as he can reach under the laws of nature, are man's rightful field of activity," and (3) "By expanding through the universe, man fulfills his destiny as an element of life, endowed with the power of reason and the wisdom of the moral law within himself."[11]

Ehricke's "Extraterrestrial Imperative" influenced space boosters to follow, as has another prescription for cosmic expansion that might be termed "The Reverse Kardashev." Soviet scientist Nikolai Kardashev developed the Kardashev Scale in 1961 as a system for detecting extraterrestrial civilizations. His scale categorized three stages of technological advance: a Type 1 civilization, or planetary civilization, like the one ours is approaching (Carl Sagan dubbed ours a Type 0.7 civilization in 1973) that can convert all the sunlight that falls on the planet into energy. A Type II civilization, or stellar civilization, harnesses the entire energy output of a star. A Type III, or galactic civilization, extracts every kilowatt of energy from an entire galaxy.

Efforts have been made to use this scale to detect stellar civilizations. Astronomers search for stars whose brightness measurements fluctuate beyond the norms, suggesting objects, perhaps even orbiting megastructures or "swarm spheres" are interfering. In 2018 there was interest in star VVV-WIT-07; another that raised interest of possible alien megastructures (interfering with light output) in 2015 was KIC 8462852, or Tabby's Star. After further study scientists concluded in 2017 that obstruction from dust clouds most likely was causing the periodic dimming.[12]

But many space advocates used the Kardashev Scale not so much as a means for searching out extraterrestrial life but to bemoan our place on the grand hierarchy. It became a motivational tool. We did not need merely to detect putative alien species and their level of development, we needed to emulate them and get past Type 0.7 to Type I, pronto, then on to Type II, harnessing the entire energy output of our Sun, and—why stop there?—Type III beckoned. If unchallenged, and fortified with plenty of Tang and Space Food Sticks, we could be rulers of the Milky Way.

Traditionally, of course, these are goals that Igor-assisted scientists set for themselves in low budget movies with "First we take the solar system, then the entire galaxy (you can keep the black holes)" speeches. They are, however, a good argument for having stayed awake during high school physics, to better parse formula-laden articles such as William A. Gale and Gregg Edwards's 1979 "Models of Long Range Growth." In the article, these physicists test out different extra-planetary civilization expansion models based on parameters that could inhibit or stimulate growth of Type II and Type III civilizations. They distinguish, for example, between "single center" growth—literally from one star—versus "multicenter" growth. They also posit, as options, a "communicative model" that assumes life cannot expand beyond a solar system, as well as a "low impact expansive model" that presumes interstellar travel is possible but rejects the possibility of the swarm spheres or the giant "megastructures" surrounding stars and harnessing their energy that Dyson and Karadashev envisioned.

Gale and Edwards then plow through calculations to establish that "the physical resources of the solar system might support a sextillion humans. This development would require five millennia at ½% per year growth. Interstellar colonization may spread through the galaxy at about a tenth the speed of light."[13] Their conclusion: according to the "single center" model we can take over the Milky Way in one million years. But, with a multicenter approach that time can be reduced: "The process of taming a galaxy takes considerable time, though it might be as short as a thousand years (by expanding simultaneously from a thousand well-placed landing points)."[14]

While the Reverse Kardashev provides a helpful expansion model, the reason for employing it, i.e., normalizing mad science, is less clear. It is a truism that science is not about "why" but "how." In the space-faring community, plenty of attention is placed on the "how" (robotics, photon propulsion, mass drivers, intergeneration starships, miniaturized humans, AI, fusion-powered skyscrapers, genetic engineering of hardy space dwellers), but, as the goals are so radical, the "why" has not been ignored (Earth's potential woes, the innate desire to explore). The philosophical answers to "why" tend to be variations on Ehricke's "Imperative"—teleological (i.e., purpose directed) explanations for why humans, or life, or intelligence, are meant to expand through the cosmos. These explanations take on coloring from the blandly scientific to the mythopoeic or unashamedly religious. The short answer: "because."

While there are a number of rationales for spacefaring, there is no firm rationality underlying the project. Like Wile E. Coyote walking off the edge of the cliff to lightly tread air, committed space enthusiasts work without a net. Arthur C. Clarke, who wrote the screenplay for 2001, conceded this point in his long essay "The Challenge of the Spaceship" (1959). Trying to answer "why?" he notes that some people are compelled to compose music, or paint and capture "a pattern of clouds that, through all eternity, will never come again," or pursue knowledge or beauty through mathematics or philosophy. He proposes, "Even if we could learn nothing in space that our instruments would not already tell us, we should go there just the same." Why? Because people like him simply "had to." They had to follow a human impulse "that needs, and can receive, no further justification than its own existence."[15]

The impulse to explore and settle space has some justification, if not in rationality, then in religious or mythic reasoning. The stars have always been associated with the heavens, the place of the gods; journeys to the stars, like the mystic's ascent described in vision literature, offer a metaphysical allure. Clarke, one of the golden age's technically versed science fiction writers, for example, cloaked space exploration in mysticism. His 1953 novel *Childhood's End* depicts an apocalyptic end to humanity's reign on Earth. When the world is at the brink of nuclear war, spaceships of seemingly benevolent Overlords appear over major cities, enforcing a reign of peace and prosperity.

The Overlords isolate human children born with paranormal abilities; these children conduct experiments that have cosmic repercussions, such as altering the Moon's orbit. These children's minds telepathically merge as they prepare to leave the Earth. Humanity, it turns out, has a destiny, born of its destruction. The Overlord's main representative, Karellen, tells the U.N.'s secretary-general "the stars are not for man"—at least in human form. The children leave their bodies and evolve into a new being, a column of light that melds with the cosmic Overmind the Overlords serve. After the children depart, the Earth itself evaporates.

*Childhood's End* with its notion of humanity's evolution into a new spiritualized species had some parallels with the philosophy of the Jesuit priest and scientist Pierre Teilhard de Chardin, whose controversial writings were published after his death in 1955. Teilhard had attended the lectures of cosmist Vladimir Vernadsky. Vernadsky popularized the idea that the Earth had gone through three phases, the first involved the geosphere, in which geological forces shape the planet, the second was the biosphere, in which biological life became a force shaping geology, and he posited as a third phase the noosphere, in which mind would become a driver of planetary change.

Teilhard, who, as a paleontologist was involved in overseeing the excavation of Peking Man in the 1920s, believed that the shaping of a noosphere, or the combined force of human minds, indicated a coming spiritual unity. He, however, rejected the model of spirituality as an ascending motion that required practitioners to deny worldliness or "terrestrial values." The material realm did not have to be denied as it was part of the spiritual process via the "cephalization of evolution." He argued that evolution, or "directed cosmogenesis," involved increasing complexity and that the advance of human intelligence conformed with a principle active in the universe. He proposed a "universe that is moving, principally and specifically, towards states of super organization" leading to "an increase of psychic interiorization and centration."[16]

Forces were pulling humanity toward cosmic enlightenment, as consciousness approached the "Omega Point" that Teilhard identified with Christ or the logos. Evolution, then, fused with the Christian notion of salvation. For his efforts, church officials issued a warning and forbade Teilhard to publish his philosophical writings. His punishment was mild compared to that of Giordano Bruno, who was burned at the stake in 1600 for supporting the Copernican (heliocentric) model of the solar system and proposing that there were numerous other inhabited worlds; it was also milder than that of Galileo, threatened with torture and placed under house arrest in 1633 for his studies that corroborated the Copernican model. Nevertheless, it was not a small matter when Church officials demanded in 1925 that Teilhard sign "a document assenting to six propositions, including traditional teachings on original sin and the place of Adam and Eve in Church teaching."[17] Although Galileo was absolved of all wrongdoing in 1992, the Church's "warning" to Teilhard, which remained as of 2019, was based on discomfort at his attempt to reconcile religion with the ungodly notion of evolution.

Although Teilhard was not a space enthusiast, the notion of increased consciousness as the goal of evolution, one way of interpreting Clarke's *Childhood's End*, has suffused the writings of many spacefaring evangelists. A case in point is Marshall Savage's *The Millennial Project: Colonizing the Galaxy in Eight Easy Steps* (1992), with Arthur C. Clarke's introduction as endorsement. Throughout, Savage passionately argues that it is time to storm heaven since we are "The sole source of consciousness in an otherwise dead cosmos."[18]

Savage adds poetry to Ehricke's dry "Extraterrestrial Imperative." We are gods-in-training, and the Age of Aquarius, of grand transformation, is upon us. Savage announces, "Now is the watershed of cosmic history . . . Life—the ultimate experiment—will either explode into space and engulf the star-clouds in a firestorm of children, trees, and butterfly wings; or Life will fail, fizzle, and gutter out . . . If we take up the sacred fire, and stride forth into space as the torchbearers of life, this universe will be aborning."[19] Savage goes on to insist that "the same unleashed powers that enable us to enliven the universe are now, ironically, causing us to destroy the Earth. . . . Thus far we have failed to use our new powers for the ends they were intended." Likening human population growth to that of a yeast culture in a bottle, Savage doesn't fret, noting instead, the "obvious answer is to blow the lid off the bottle!"[20] As in Ehricke's "Three Laws," and Clarke's *Childhood's End*, nothing can stop us from leaving our two-bit playpen of a planet.

Savage's eight step program includes first creating "Aquarius," floating cities on hexagonal platforms whose inhabitants can farm algae, seaweed, shellfish, and fish, and create electricity via heat engines. In fine detail, he describes these floating islands with their breakwaters, greenhouses, parks, and underwater apartments.[21] The next step in Savage's plan, or "Asgard," involves building space colonies. Savage goes on to describe a Moon colony "Avallon" (a domed crater), then the terraforming and settling of Mars or "Elysium," followed by asteroid settlements, and "Solaria" or working to create a Dyson sphere (or a "Dyson swarm" around a star) to boost humanity to a Type II Kardashev civilization. When the human population of our solar system approaches the trillion mark, with 95 percent of humans living in the asteroid belt, it will be time for further expansion. Savage depicts a Type III civilization spread through the galaxy. Like a true cosmist he argues that exponential growth and spread of the species "will allow us to annihilate annihilation itself."[22]

Savage concludes his work with a call to form the First Millennial Foundation to begin the great enterprise, as readers could not wait for governmental action. He urges the development of "a network of free individuals," a group whose "core members are sacred warriors in the Cosmic struggle for the universe."[23] Members of the Foundation, named after the Isaac Asimov novel series, will be a "tiny handful of the god-beings who inhabit the universe."[24]

Savage's inspired call to arms failed to create a mass movement but did attract supporters. Carl Sagan reported receiving a mailing from the group urging a contribution of $120 yearly to ensure "citizenship in the first space colonies—when the time comes."[25] After quarrelling with Savage, a group of followers created the offshoot Living Universe Foundation. These enthusiasts decided to develop the Space Environments Ecovillage (SEE) on twenty acres in Bastrop, Texas (at 135 Millennial Way).

William Gale, the Bell Labs physicist who coauthored the previously quoted article on "Models of Long Range Growth," bought and donated the twenty-acre site to the SEE. While Gale remained on the East Coast, the first settlers arrived in 2001. At its founding one settler noted: "While we are ordinary people, we do have a grand dream. If we are to help spread life throughout the galaxy, which *must* be done sustainably, we need to start now . . . We do not have the funding to do massive technical research, but we can work on the basic problems of living in a remote environment."[26]

Ultimately, a collection of homesteaders in trailer homes did not add up to an Aquarius or Asgard. Like many a utopian colony before it, the ecovillage went defunct in 2004. On the brighter side, Savage helped develop a board game that an Australian company released in 1999. The product, "6 billion," for ages twelve and up, was a "game which combines the themes of exponential population growth with the colonisation of our Solar System."[27] Reviews indicate that some gamers remained fond of it as of 2019, although one noted "the theme is just far too optimistic." Also of note, now that seas are rising, glaciers are melting, and major flooding of coastal cities is inevitable, variants on Savage's seasteading plan have made a comeback. In April 2019, for example, the United Nations Human Settlements Programme presented a round table to discuss Oceanix, a Dutch designer's plans for floating cities, made of linked 4.5 acre hexagonal platforms, each of which could house 350 people.[28] It is possible that Savage really was a prophet, sketching out, in inspired fashion, a distant future; his dreams of Aquarius and Asgard emerging, perhaps, after hundreds of years.

While Savage insisted his Millennial Foundation was not to be mistaken for a religion, it offered a template for forging utopian communities in a purposive cosmos that awaited a transformed humanity to "kindle the green flames of a billion billion worlds."[29] He offered technical details wrapped in late twentieth century New Age millennialism. Of the foundation, he stated, "It is certainly not a church or a religion, although it has its roots in our numinous relationship with the cosmos."[30] Expansion into space was our duty. He was certain that humanity had an important relationship with the universe; Savage asserted that as there are unlikely any other sentient extraterrestrial species, "our responsibility to the Cosmos is absolute."[31]

• • •

This kind of talk has made some people uneasy. One of them was the director of the Vatican Observatory, Guy Consolmagno, a Jesuit brother and research astronomer. I asked him whether the idea that humanity had a destiny in space was widespread among astronomers, or existed in a small echo chamber. He responded, perhaps fittingly, on April Fool's Day, "The premise, while true I am sure of many people including those you cite, really isn't the way that anyone I know of thinks about the exploration of space. I don't see it as any sort of destiny; it's much more matter-of-fact. And in some ways I am suspicious of those who turn space exploration into a religion."[32]

A variation on this theme was sounded by Rusty Schweickart, an Apollo 9 astronaut, and one of many who reported a mystical experience while in outer space. He rejected the tendency to insist humanity had a destiny to expand and fill the universe with life or mind. "The Earth has been a womb for the development and evolution of intelligence, and I think that it is undergoing a very basic and radical change now. That is not a motivation—that is just an observation. I think you have to be very, very careful talking about things of this kind because people turn it into a religion, and it becomes a purpose instead of an 'awesome observation' . . . that element, where it moves away from awe and becomes a "purpose" is a shadow side of religion that makes me very uncomfortable. Whether it is in the form of space religion, or traditional God religion, or communism, or what have you."[33] While Schweickart agreed the emergence of consciousness was "awesome" he did not want it to undergird a mythopoeic vision of the cosmos—or one of humans as gods-in-training. He was concerned about fanaticism in any movement, and the failure to recognize the yin-yang or "shadow and a light side of every element of creation."

A harsher verdict came from C. S. Lewis. A devout Catholic, Lewis blasted the spacefaring heresy when he discovered it in its infancy prior to World War Two. His critique could be summarized by adding the judgment "this is bad" to Sigmund Freud's statement that mankind, via technology, was making itself into "prosthetic gods." The "gods in waiting" theme, to Lewis, was noxious. Lewis's novel *Perelandra* (1944) critiqued Interplanetary Societies and Rocketry Clubs and the idea they were based on, which was likely to "open a new chapter of misery for the universe. It is the idea that humanity, having now sufficiently corrupted the planet where it arose, must at all costs contrive to seed itself over a larger area." He continued, likening the idea to a masturbatory fantasy, "But beyond this lies the sweet poison . . . the wild dream that planet after planet, system after system, in the end galaxy after galaxy, can be forced to sustain, everywhere and for ever, the sort of life which is contained in the loins of our species—a dream begotten by the hatred of death upon the fear of true immortality, fondled in secret by thousands of ignorant men and hundreds who are not ignorant." Clearly, Lewis did not interpret scripture in accord with the equally devout Federov, who longed to make humans immortal, resurrect all the dead, and encourage spacefaring.

And yet, the notion that the universe has a built-in script with a starring role for sentient beings, if easy to critique, is hard to shake, and can find some shelter in recent cosmological theory. This inroad involves the notions of the "weak" and "strong" anthropic principle. The weak principle has a simple premise, which Stephen Hawking and a coauthor explained: "The fact of our being restricts the characteristics of the kind of environment in which we find ourselves."[34] Simply put, life, and consciousness, can only emerge in an environment conducive to it. So the fact that we are on just the right kind of planet, just the right distance from a star of a certain category, with the axis tilted just so, with plenty of surface water, an appropriate atmosphere, and so on is not really improbable, in the sense that such an environment would be required for our emergence.

However the strong anthropic principle doesn't only suggest bacteria and lungfish and apes appeared on the right planet, but that the universe at its very basis is designed to conjure up sentient life—with consciousness the ultimate dance partner for matter. Astrophysicist Fred Hoyle, who, in the 1950s pioneered the idea that elements were formed in stars, sketched out an early brief for what later became known as the strong anthropic principle. He argued that one of the most improbable factors of our universe was the emergence of a stable form of carbon in vast quantities from nuclear fusion reactions in stars. The unlikely "triple alpha process" that formed carbon required three helium nuclei to fuse to create carbon atoms. Hoyle puzzled over why there was so much carbon in the universe when this process should have yielded only small quantities. He predicted that carbon, in a certain quantum state, would have to match "the sum of the energies of a beryllium nucleus and a helium nucleus," to induce a resonance that would help propagate the triple alpha process.

This prediction was later confirmed and Hoyle remarked, "I do not believe that any scientist who examined the evidence would fail to draw the inference that the laws of nuclear physics have been deliberately designed with regard to the consequences they produce inside the stars."[35] Analyzing other "coincidences" in the universe's make up, Hoyle added, "Rather than accept a probability less than 1 in $10^{40000}$ of life having arisen through the 'blind' forces of nature, it seems better to suppose that the origin of life was a deliberate intellectual act."[36] He also quipped that the universe seemed a "put-up" job, and that the "carbon atom was a fix."[37]

Observations similar to that of Hoyle led in the 1980s to discussion of the universe's built-in "fine-tuning" to make the conditions for life possible. If, on some large control panel, various basic constants found in nature had been altered slightly, we would not be here to discuss them. For example, Einstein introduced the "cosmological constant" in 1917 to maintain the model of a static universe in which gravitational forces would not lead to the universe's collapse. This constant represents the energy found in "empty" space—more recently identified as dark energy—that counteracts gravity.

A decade after Einstein plugged the cosmological constant into his equations, Edwin Hubble analyzed the light spectrums of distant stars and determined the universe was not static but expanding. Physicists have since shown that a large value for the cosmological constant would accelerate cosmic expansion and stars and galaxies would never have formed, while a negative value would cause the universe to collapse. The gravitational constant, as well, hovers at a Goldilocks point—high enough to permit the existence of stars and their fusion reactions, but not so high that our star's "lifetime" would be too short to allow the emergence of life.

Proponents of the strong anthropic principle would suggest our universe was "rigged" to have been favorable to the emergence of life and intelligence. Theoretical physicist Paul C. W. Davies argues this isn't as outlandish as it sounds. He noted that it was once common for the science-minded to think of life as a "trivial embellishment" to the physical world, unimportant to the larger workings of the cosmos. For example, novelist Kurt Vonnegut Jr. wrote, "I tend to think of human beings as huge, rubbery test tubes, too, with chemical reactions seething inside."[38] Similarly, Stephen Hawking—possibly on a bad day—stated during a 1995 radio interview that "The human race is just a chemical scum on a moderate-sized planet, orbiting around a very average star." But, in contrast, Davies insisted "conscious organisms should not be casually shrugged aside as just another sort of physical system." Instead, he argued that "life is a key part of the evolution of the universe" and that "mind is linked into the deep workings of the cosmos."[39]

Davies assessed consciousness as too precious to be regarded as a by-product; he believed it might be a "fundamental property of the world." The role of the observer in quantum mechanical interactions, he argued, suggested the intertwining of consciousness with the most elemental processes of the universe. Conclusion: as Marshall Savage would put it, we have a responsibility to spread life and intelligence through the cosmos. Two cosmologists, John D. Barrow and Frank J. Tipler, in their 1986 book *The Anthropic Cosmological Principle* argued that spreading mind through all the universes ultimately would result in Teilhard's Omega Point.

Most physicists draw different conclusions. Stephen Hawking, to give one prime example, found a large loophole in anthropic reasoning—in fact, our universe was the loophole. In *The Grand Design*, rather than endorse some version of intelligent design, Hawking and coauthor Leonard Mlodinow backed a concept straight from the pages of science fiction—that of the multiverse. They based their endorsement on the confluence of quantum theory, string theory, and Alan Guth and his colleagues' development of "inflationary theory" in the 1980s.

According to some versions of inflationary theory, there are multiple universes. The "Big Bang" was a local event, part of an endless, eternally expanding cosmos, with big bangs spawning infinite disconnected bubble universes from the quantum flux. String theory posits that each of these universes can have different laws of physics and constants. Ours is just one of a near infinite set of these bubble universes. One justification Hawking presented springs from Richard Feynman's theory of "sum over histories" that indicates for a subatomic particle to get from point A to B, it doesn't take only the direct path but every possible path, including one around Saturn and back. Relating inflation to quantum theory, Hawking wrote, "The universe appeared spontaneously, starting off in every possible way. Most of these correspond to other universes."[40] Cosmic inflation is not the only theory that leads to the notion of the multiverse. Physicists Raphael Bousso and Joseph Polchinski calculated that string theory has on the order of $10^{500}$ solutions, each corresponding to a different possible universe.

The concept of the multiverse turns the strong anthropic principle back into the weak anthropic principle. If there are an infinite number of universes, or the more modest figure of $10^{500}$ from string theory that Hawking and Mlodinow relied on, each with unique laws and constants, then our universe is no longer unique. It just happens to belong to the subset of bubble universes that can support life. And if the number of universes is infinite, just as monkeys, given infinite time, and lots of laptop batteries, could type out Dr. Seuss's *Green Eggs and Ham*, at least one of these other universes may be an exact duplicate of ours. And in that universe, as you thumb through this work, you are contemplating whether to start your own space news podcast.

Many prominent contemporary astrophysicists have supported the multiverse concept, including Steven Weinberg, Alan Guth, Martin Rees, Leonard Susskind, and Hawking. In 1987, Weinberg used the underlying anthropic principle to determine the cosmological constant's value, and a decade later his value was experimentally confirmed. However, despite this feat, Hawking acknowledged that the anthropic principle was not popular among all physicists, nor, it seemed, was the multiverse theory. The concept of the multiverse has stirred doubt. On aesthetic grounds, Freeman Dyson found it silly, "I think the multiverse idea doesn't make much sense. There is only one universe we can deal with. The rest is fantasy."[41] Sabine Hossenfelder, a theoretical physicist at the Frankfurt Institute for Advanced Studies has commented, "Nobody who does serious science works with the multiverse because it's utterly useless."[42] Even Hawking had his doubts. Several years after writing *The Grand Design*, he remarked, "I have never been a fan of the multiverse." A paper by Hawking and an associate published after Hawking's death did not discard the theory but noted, "Our findings imply a significant reduction of the multiverse, to a much smaller range of possible universes."[43]

• • •

To clear some of my confusions on these matters, I visited the office of my friendly neighborhood astronomy professor, Erin O'Connor at Santa Barbara City College. The campus's white buildings were perched above a remarkable view of sparkling blue sea. It was late in the afternoon and he still had two of his students in his office. I pulled in an extra plastic chair. Professor O'Connor was a large man of boundless enthusiasm, a shade over fifty, with a light moustache, a Hawaiian shirt, and sandals with socks. The walls of his office were covered with artwork, including a poster depicting subatomic nuclear reactions, a framed photograph of astronauts, a line-up of NASA, Roscosmos, and billionaire rockets, and a pen and ink drawing of Kirk and Spock of the classic *Star Trek* television series with a piece of tape above Spock's mouth where a student had drawn an O'Connor-esque moustache. He also had another sketch, of, presumably, Spock's hand in the "Live Long and Prosper" double-finger scissors spread.

He encouraged his students to remain (and check their phones) while I fired questions his way: Were humans destined to go into outer space? O'Connor, seldom at a loss for words, declared that a case could be made for that from evolutionary biology. Evolutionary pressures provide a push; people are wired to do crazy stupid things—not all, but enough to strike out into the unknown now and again. Someday we certainly would go to space—there would be viable colonies on Mars. But we were talking thousands of years in the future, and O'Connor believed that within 100,000 years it was almost inevitable. Was the emergence of intelligence a "cosmic" event? After reminding his listeners that he had no standing, he suggested that not all cosmologists would adhere to that, but it was a reasonable assertion and generally accepted that intelligent life is a "special" form of matter.

How about the multiverse? I asked. Was it popular among astronomers? Well, how crazy was it, he asked, that everything was just right so we could ask "Why are we here?" He confessed he once had thought the multiverse theory "wacko" until cosmologist Robert Geller of UCSB convinced him that since the probability for our universe was nearly zero, the multiverse theory made a case for it as one of the universes that supported life in an infinite set of possible universes. His students and I then got into a brief discussion about science fiction, and Professor O'Connor left me with the helpful reminder: "remember that scientific theories can be mechanisms to get answers, not necessarily to provide descriptions."

Next, to test how well the multiverse theory was accepted, I checked in with the podcasts of Brother Guy Consolmagno. He definitely objected to using cosmological theories as mechanisms for retrieving answers on theological matters. This effort, of course is outside the purview of most physicists and cosmologists, but one he ran across. Because of his role as director of the Vatican Observatory—*the Vatican, where Galileo was put on trial for using a telescope!*—he was sought out at times to reconcile religion and science. That was a task that didn't interest him. He hewed to the "separate spheres" approach to these realms. "Mixing" could be toxic.

Regarding the anthropic principle, he cautioned fellow believers against leaping on the fine tuning argument as a way to prove the existence of God. Such proofs disturbed him as they began with science and put God at the end of the chain of reasoning. He noted, "To a scientist who's a believer, it goes the other way around."[44] He added, "The fact that we exist at all asks the question, why is there something instead of nothing . . . I'm not talking of the creation at the beginning of time, I'm talking about the fact that existence continues to exist. And that there is not only this cosmological constant or all the other fine structure constants that go into the anthropic principle. But even if there are an infinite number of multiverses with different ways of physics—and some of them ending very quickly and some of them lasting forever . . . and some of them having creatures with twenty-seven tentacles that think in totally different ways than we do. . . . all of them are the creation of God. . . . However you make the universe, God will be bigger. That is what it means to be infinite."[45]

The multiverse might prove to be valid, but wasn't necessary to his faith. And, as he'd expressed earlier, outer space need not be anyone's destiny. Clearly no cosmist, Consolmagno implied that space may not be the best form of religion. While here he was in agreement with astronaut Rusty Schweickart, this is not to say that there is no room for mixing "awe," even of a spiritual sort, with space travel. Clearly, for many space proponents, there is a sense of wonder, perhaps even of the sacred, motivating them. The insight linking microcosm to macrocosm, common in Renaissance thought, in which each person was seen as simultaneously minute, and a universe onto him or herself, is one of the secret pleasures of all star gazers, including planetarium visitors.

A fair portion of space proponents sense a sacred dimension to their obsession. As planetary scientist Wendell Mendell, a long-time space activist and lunar colony designer, noted, "I find the passion behind many peoples' interest in space to be almost religious in nature. (I am not the only one to notice this.) I have encountered many individuals or groups who have some transcendent vision of future activities or assets in space."[46] Likewise, Carl Sagan, with many caveats, noted, "I worry about people who aspire to be 'god-like.' But as for a long-term goal and a sacred project, there is one before us."[47] That would be spacefaring.

It is reasonable to assume scientists are not immune to a sense of awe from contemplating humanity's small yet seemingly significant place in the cosmic scale, as indicated in this saying of eighteenth-century

Rabbi Simcha Bunem of Pershyscha: "Each person should carry with them two slips of paper in their pockets. In one pocket should be a slip that reads 'I am but dust and ashes.' In the other should be a slip that reads 'For my sake the world was created.'"[48] Or, as Erin O'Connor, the astronomy professor at Santa Barbara City College put it, "It is not enrichening to think that humans are simply bags of chemicals. I don't see any problem in finding a deeper meaning in existence and reconciling that with science." But allowing oneself a feeling of awe, or even of humility and resolve, is different than dedicating oneself to the ultimate goal of redeeming the entire cosmos by filling it with life or consciousness. This is the mythic task set by Fedorov, the cosmist, his version of the Christian project of resurrection, bringing life from death.

The company of true believers might be divided into those who are passionate about spacefaring but subscribe to "separate spheres" for science and religion, those who connect spacefaring with a sacred mission, and those that could be said to take a gamer's mentality to the puzzle of being dropped into the vastness of the cosmos. (Regarding this latter category: one of visionary Marshall Savage's most tangible achievements thus far has been a board game about space conquest.) To such a gamer, the universe might appear as a really vast three-dimensional (or for string theorists, ten-dimensional) game board. The game will be won when a team benefits from, in some way, every atom in the universe. Many in the spaceflight echo chamber are not so much Mad Scientists out to conquer the universe, as gamers anxious to grab up all the properties, even the lowly Baltic Avenues, by sleight of hand, if possible, or by force if necessary. After all, asserting greater and greater control over brute nature is rule number one in the progress game manual. But do we rightly attempt to turn everything into a tool, or do we learn to leave things alone—even asteroids and comets otherwise devoid of life?

Olaf Stapledon, one of the early masters of science fiction, known best for his philosophical novel *Starmaker* (1937) describing a dream journey through the cosmos, proposed another version of "winning."

In his 1948 essay "Interplanetary Man?" originally a talk given to an audience that included Arthur C. Clarke and other members of the British Interplanetary Society, Stapledon warned that if humanity got past the danger of atomic annihilation it could easily "slip in to the assumption that the goal of all its corporate action is simply to make a bigger and bigger mark on a bigger and bigger environment. Power is all too apt to become a goal in itself."[49] Absent ethics, this was an example of misplaced faith.

Stapledon argued that humanity's true task has always been to come into conscious community with others. If there were extraterrestrial species, and we were to make contact, he hoped our goal would be to enter in community with them in a "Commonwealth of Worlds." If no sentient extraterrestrials were found, we might yet come into a broader community with the transhuman species that might eventually develop on our own settlements scattered through the solar system or galaxy. Conquering the universe did not appeal to Stapledon, but he, like Teilhard, approved the notion of the awakening of consciousness—or fellowship. Stapledon hoped if there was some "purpose" in the cosmos it was in the creation of a grand community, but acknowledged that because of the vast spatial distances, contact with extraterrestrial species might never occur. If that were true, he hoped that God or some vast cosmic force might embrace us all in a greater mind. Stapledon's worst-case scenario: we might be like cabbages in a garden, grown, at best, to be plucked and eaten—unable to surmise the greater forces consigning us to a limited role. Our true destiny might just be that of bit players in the cosmic scheme.

# CHAPTER ELEVEN

# THE SPACE RAVE

• • •

"Robots aside, we've backed off from the planets and stars.
I keep asking myself: Is it a failure of nerve or a sign of maturity?"
—Carl Sagan, *Pale Blue Dot*, 1994

The world's true launch pioneers, space monkeys and dogs, get some due—exhibit A: Miss Baker, the squirrel monkey who, along with her rhesus companion, Able, survived a flight in a Jupiter missile in 1959 and still has bananas laid on her gravesite in Huntsville, Alabama.[1] But they pull out all the stops, and not just in former Soviet nations, for cosmonaut Yuri Gagarin, who, on April 12, 1961, became the first human in outer space. In April 2019, I attended "Yuri's Night" a "world space party" in Los Angeles at the California Science Center. A phalanx of Imperial Storm Troopers, a futurist drag race car, replete with sinuous, spaghetti-like chrome and lit with purple and green fluorescent colors, a Mad Max hot rod, and small robotic rovers were at the entrance.

Inside, much of the museum was open to guests, including the aerospace galleries with a full-size replica Wright glider, an actual Apollo-Soyuz module, the Gemini 11 capsule, and a virtual reality rendering of the interior of the Apollo 11 Moon lander (a newbie to VR, I kept trying to grab the handles, and cetera, available in the lander). Of the 1,500 mostly young space enthusiasts attending, many were from the high-tech industry, some in silver jumpsuits rigged with blinking LED lights, others in blue NASA uniforms, Star Trek regalia, Stormtrooper carapaces, neon-bikini fuzzy-ear alien get-ups, or varying space buccaneer costumes. The party climaxed below the heat-scarred Space Shuttle Endeavour (complete with red lettering near its tail "Cut Here for Emergency Rescue") where, through glowing headphones, we listened to speeches appropriate to a revival meeting given on a stage by Bill Nye, the former Science Guy, now CEO of the Planetary Society ("Greetings ladies and gents, citizens of Earth!" he began his speech), Virgin Galactica pilot Mark Stucky, and NASA's wildman-astronaut, Story Musgrave.

Musgrave, in his machine-gun patter, told of growing up on a farm operating and fixing tractors and combines, working simultaneously for two branches of the military driving and repairing tanks and other vehicles, this while also talking his way into medical school without a high school diploma, becoming a surgeon, and gaining other graduate degrees (math, computer science, chemistry, medicine, physiology, literature, and psychology). While a NASA astronaut, he was also a trauma surgeon and a mechanic, and made seven space shuttle flights. Prepped by working on a replica, his crowning effort, taking a spacewalk in 1993 off the Space Shuttle Endeavour—above us right there in the hall!—to fix the problem-plagued Hubble Telescope (because of its decaying orbit, the Hubble now was expected to crash into the Earth in the 2030s).

The crowd, which had been punctuating his litany of triumphs with rebel yells, went wild. He seemed like Gully Foyle of Alfred Bester's *The Stars My Destination* (1956), a rough and ready, semi-rogue genius and man of action. This was a guy whose every cell participated in the power of positive thinking. Musgrave's goal was to inspire his youthful audience to break down all barriers in pursuing dreams and left us with the thought that we were on a "quest for the cosmos, quest for the universe, quest for who we are as human beings."

Aerospace companies, high-tech startups, and research centers were there: JPL, Virgin Galactic, the Spaceship Company, Aerojet,

Rocketdyne, Axiom, as well as the Planetary Society (the love child of Carl Sagan), and the National Space Society (Wernher von Braun's), also university departments (including my friends from UCSB's Experimental Cosmology group). As I wandered, more than once I came upon people with cell phones videotaping friends who were giving "raps" about the need for science, with hand gestures and dance moves. As should be no surprise when wandering a museum filled with people dressed in NASA and Star Trek uniforms, there were plenty of people who believed in the star settlement ideal. What exactly dancing to a live deejay's hip-hop feed through headphones while wandering a tunnel flanked with an aquarium's well-populated coral reef had to do with space is anyone's guess—perhaps we were modeling what a Saturday night on a space station or lunar lava tube might be like—but fun nonetheless.

These were not typical Americans. If they were typical Americans, only 35 percent would believe in human evolution "due to natural processes," while 24 percent would believe a "Supreme being guided evolution," and 31 percent would believe that humans "existed as is since beginning." Likewise, if typical Americans, more than half of them, 52 percent would believe that scientists are unclear on whether there was a big bang.[2] And, more alarming, 6 percent of them would report the Apollo 11 landing was staged, while 15 percent would be unsure if it was staged or not.[3]

While the silver-suited aliens around me were not typical Americans, most were millennials, so at least 63 percent of them would be interested in space tourism, while of their older peers, the Gen Xers, only 39 percent would want to be space tourists, while people in my demographic, Baby Boomers, were 27 percent interested. (I also am about 27 percent interested in space tourism.)[4] Of greater concern to spacefarers, only 32 percent of typical Americans think that within fifty years, that is, by 2068, humans will build colonies on other planets that can be lived in for long periods; 67 percent believe this will not happen. Much to the credit of common sense, in 2018, most typical Americans thought that NASA's goals should be to monitor the Earth to aid in climate research (63 percent call it a top priority) and monitor asteroids that might hit the Earth (62 percent labeled it top priority). This was a much larger percentage than those who thought there was a strong need to travel to Mars (18 percent) or return to the Moon (13 percent). Clearly, from all this data, the main message (remember this was spring of 2019): no one but NASA and the Trump administration—particularly Vice President Mike Pence, who was all about returning to the "heavens"—really wanted to go back to the Moon.

Now, was I mulling this data while wandering through the Science Center, trying not to photo bomb rapping science vloggers and passing bars with $10 beers and even more expensive mixed drinks? No. I made to exit the Endeavour Hall when New Age music came on and people were asked to hold hands and try an experiment. "Imagine you have stepped onto another planet of compassion," said the speaker. Soon people in blue and orange NASA uniforms were linking hands with Klingons, people in tutus and spandex and pipe cleaner antennae, and others wearing flashing LED lights, and it was really hard to get to the free burrito line. I was trying to disprove Robert Heinlein's roughly translated maxim TANSTAAFB, or "there ain't no such thing as a free burrito." The hundreds involved in the New Age exercise were now told to whisper a wish or blessing into their neighbor's ear.

I went off to try to find some likely space enthusiasts to interview and stopped at an exhibitor with NASA memorabilia. "This flag has flown on Apollo missions," I was told. Near it were hermetically sealed food rations also dating to Apollo days. Without the labels, the meals were impossible to identify. "Yummy, huh?" asked the exhibitor. I wandered to another area where space artists had set up displays. Marilynn Flynn's paintings were impressive. Her book, *Space Art*, included a highly detailed landscape of "Balgatan Breach—Europa," with its shattered cliffs, and a view of Jupiter, as if it were the moon, looming above the horizon, while water vapor shot from crevasses. Her view of Jingpu Rivas on Titan, the moon of Saturn, featured methane streaming down hillsides to a turbulent lake of liquid methane. Flynn, an Arizona native, had been part of a NASA fine arts in space program—her work to be displayed on the Challenger Space Shuttle that exploded in 1986. She was a classic "orphan of Apollo" as older enthusiasts are referred to. We would, she said, definitely one day be spacefarers. "We are a migrating species, we started in Africa, and I think we're trying to migrate into space. It's just embedded in our brains." She sells her work at space conferences and conventions but has given up on science fiction conferences. Tired of the furries and other silliness.

Down from her, a fellow named Rex, wearing a home-made Star Trek uniform, and an amateur space historian with the goal of hosting a space radio program, behind a table stacked with old newspapers with headlines chronicling the journeys of John Glenn and others, was even more expansive. We would not only settle the solar system but the stars. "A warp drive is science fiction to us now—" he leaned over to a woman who was beginning to pick up the yellowed newspapers, "excuse me, but those are really delicate."

"Can I look through them?"

"They are very fragile."

Rex continued, "You go upstairs and you see the Wright Brothers' glider. Only sixty-six years after they built it we were going to the Moon." Rex, I soon discovered, was well-versed in the technical aspects of rocketry; he was planning a book that described every mission of the space programs of both the U.S. and U.S.S.R. He continued a rap with which I was getting familiar. It referred to the Earth's growing population. The adventuring pioneering spirit to be found in Americans. Time to step out into space.

"Who was your favorite Star Trek captain?" I ask him. (This was not a non sequitur; Rex was wearing a home-sewn Star Trek uniform.)

"No question about it. Benjamin Sisko from *Deep Space Nine.*"

I told him Sisko started out great but somewhere in the series it seemed like he'd gone through an emotional trauma and was just ghost walking the part. He no longer seemed Shakespearian or commanding.

"That's nonsense," said Rex. "His acting was superb throughout all the seasons. Avery Brooks was also excellent in the television show *Archer.*"

He had his ideas. I had mine. Probably he was right.

I wandered into a room where partygoers were using scoops to try to pull balls up an incline. Each scoop represented a different type of bird beak, and success or failure indicated the limits of its evolutionary design. It reminded me of the "Shoot the Moon" game we used to play at my grandfather's apartment when I was a kid. The premise was to defeat gravity by manipulating two metal rods set in a kind of "V" pattern on an incline to coax a large metal ball uphill; depending on your ability the ball would drop in the first slot: Earth, or, in ascending order, First Stage, Second Stage, Third Stage, Fourth Stage, and, at the top of the scale, the Moon. The trick, as I recall was to get the ball to drop a bit and then "squirt" up in a burst—or perhaps a series of these squirts. Let's call this a metaphor for becoming a spacefaring species. As the Moon Shot was first built in the 1940s the sights were low. Today's game was, well, intergalactic travel. Was it possible? Or were we stuck at "Moon" or "Mars?"

I had been circulating through an echo chamber where everyone was determined to Shoot the Stars. When the night's deejay had asked "How Many of You Want to Go into Space!" virtually every hand was raised along with a loud roar. Even outside the echo chamber, interest was mounting. In a 2015 poll, even 47 percent of the members of the American Academy of Sciences agreed that humans were essential to the future of space exploration. Well over half of millennials, or young adults, were interested in space tourism—which many strategists consider, along with satellites and asteroid mining, a reasonable step to building space infrastructure. To make it all happen, Elon Musk was then completing SpaceX's private launchpad in Texas near the Rio Grande and beginning tests on the BFR, since renamed the "Starship," that could carry one hundred colonists. Were we to become spacefarers? Would we have functioning space settlements in a few centuries or millennia? And would this be an achievement—or what Ken Kesey had deemed "a turning away from the juiciness of stuff?"

That was the sort of thing I was mulling as I wandered past the Virgin Galactic table to the adjacent table for "The Spaceship Company." I liked the name; it reminded me of the Acme Company that Wile E. Coyote had once depended on for buying Road Runner-defeating equipment. The Spaceship Company was, as I learned, the name of the Virgin subsidiary that, yes, made actual spaceships, more exactly, ships designed for Virgin Galactic's impending suborbital flights.

The Spaceship Company's young and energetic employees stood behind the table alongside those for Virgin Galactic and Virgin Orbit. Many looked ready for a high fashion photo shoot. I began talking to Christian Engelbrecht, an avionics (electronic systems) test engineer, who appeared athletic and cheerful. He told me that he was a fan of SpaceX, since they were not direct competitors—SpaceX's tourism ideas, taking passengers around the Moon, and so on, put them in another league. He wished them all the success. He added his company had a "friendly" rivalry with Blue Origin, as they competed for the luxury suborbital flight market—and yes, commercial spaceflight was imminent. The Spaceship Company would remain near Edwards Air Force base in Mojave, California, but Virgin Galactic was just then moving its headquarters from Mojave to the Spaceport America in New Mexico.

I sketched out for him my interest in people who believed humanity had a destiny in space. That it was our responsibility as a species to enliven and populate the cosmos. Did he agree with this? Did he meet people with this sort of mindset? He thought for a moment then said, "People can feel very strongly, in a religious sense, these ideas about space. But I don't think there is a fixed 'destiny.' That we have to go. I would take the existential route to make that decision."

"It's something for people to decide?" I was in shock over the simplicity of his remark.

"Exactly."

He had homed in on a problem I had with spacefaring enthusiasts. I admired their enthusiasm, their obsessions, brilliance, and borderline wackiness. I like wacky. I admired that they were dedicated to goals that might take centuries or millennia—if ever—to achieve. But I sensed that by laying claim to humanity's inherent "responsibility" to the cosmos, true believers were ducking real responsibility. It was important to recognize, as with Feynman's sum over all possibilities, that we had choices for pathways—perhaps not infinite—but more than one. Even William A. Gale and Gregg Edwards's article "Limits to Growth" factored in civilizations that chose not to or that could not leave their solar systems. While tracing out the possible paths for a high-science future we also should factor in sanity. We, as a species, have to make our own choices, accept their full weight, and not simply the assurance that salvation is to be found in the stars.

# ACKNOWLEDGMENTS

• • •

Writing this book entailed stepping out onto the thin ice of my understanding of space science—to maintain the proper altitude required the knowledge, advice, and musings of many colleagues, relatives, friends, and space experts. My contribution, I hope, will be to add to the history of spaceflight a glimpse of the cultural cauldron in which it continues to bubble.

I have many people to thank. Taylor Dark III not only allowed me to portray his interest in space colonies as a teenager, but read and commented on many of the chapters. My brother Steve Nadis, a science journalist and author, provided a crash lesson on cosmology. Alan Shikami and Saul Nadis, my son, read drafts and commented on chapters. Jatila van der Veen, a project scientist at the University of California at Santa Barbara (UCSB), offered advice and arranged meetings with members of UCSB's Experimental Cosmology Group as well as the undergraduate team competing in NASA's Big Idea Challenge 2020.

Those who generously shared their ideas include Philip Lubin at UCSB, Freeman Dyson of the Institute for Advanced Study, physicist James Benford, Robert Zubrin of the Mars Society, Lisa Spence at NASA, John Spencer of the Space Tourism Society, Erin O'Connor of Santa Barbara City College, Smithsonian planetary scientists Bruce Campbell and Jim Zimbelman, Dale Skran of the National Space Society, and authors Kim Stanley Robinson and Andy Weir. I interviewed many other members of the space community and send my appreciation to all. The archivists at the Smithsonian Institution's National Air and

Space Museum helped me pack much research into a short visit. Maxximilian Seijo contributed translations from German. My gratitude also
to Jeffrey Meikle, Bruce Hunt, and Michael Saler, all generous allies.

The genesis of this project, in part, came from childhood viewings
of the science fiction puppet series *Fireball XL-5* ("filmed in Super-
marionation"), as well as discussions about the spaceflight idea with
Taylor Dark III when we were colleagues at Doshisha University in
Kyoto, and with Ryan McMillen while in graduate school at the University of Texas at Austin. I am grateful to the Alfred P. Sloan Foundation for awarding me a fellowship in their Public Understanding of
Science and Technology program to complete this book in 2019–20.

Many thanks to my wife Kate Connell, who kindly accepted and at
times aided and abetted my two year submersion in all things space-y
and to my daughter, Rose Nadis, for working a reference to Mars
settlement into one of her songs. My gratitude, as well, to my agent
Mark Gottlieb at Trident Media and my publisher Claiborne Hancock
at Pegasus for their enthusiastic support for this project.

# ENDNOTES

· · ·

## PREFACE

1 David Ketterer, *New World for Old: The Apocalyptic Imagination, Science Fiction, and American Literature* (New York: Anchor Press, 1974).

2 Richard J. Gott, "Longevity of the Human Spaceflight Program," in Edward Belbruno, ed., *New Trends in Astrodynamics and Applications III, American Institute of Physics Conference Proceedings, Volume 886,* (Melville, NY: AIP, 2007).

3 Tom Wolfe, "One Giant Leap to Nowhere," *New York Times,* July 18, 2009, https://www.nytimes.com/2009/07/19/opinion/19wolfe.html.

4 Howard Bloom, "The Big Burp and the Multiplanetary Mandate," in Steven J. Dick and Mark L. Lupisella, eds., *Cosmos and Culture: Cultural Evolution in a Cosmic Context* (Washington, DC: NASA, 2009), 145.

## CHAPTER ONE—MARS MANIA 1.0

1 William Sheehan, *The Planet Mars: A History of Observation & Discovery* (Tucson: University of Arizona Press, 1996), 79.

2 Camille Flammarion, translated by Brian Stableford, *Lumen* (Middletown, CT: Wesleyan University Press, 2002), 56.

3 Bernadette Bensaude-Vincent and Liz Libbrecht, "A Public for Science. The Rapid Growth of Popularization in Nineteenth Century France," *Réseaux, The French Journal of Communication* 3, no. 1 (1995): 75.

4 Bensaude-Vincent and Libbrecht, "A Public for Science," 87.

5 Robert Markley, *Dying Planet: Mars in Science and the Imagination* (Durham, NC: Duke University Press, 2005), 50.

6 David Strauss, *Percival Lowell: The Culture and Science of a Boston Brahmin* (Cambridge: Harvard University Press, 2001), 178.

7 Markley, *Dying Planet,* 64.

8 Robert Crossley, *Imagining Mars: A Literary History* (Middletown, CT: Wesleyan University Press, 2011), 76.

9 *Publications of the Astronomical Society of the Pacific* 6, no. 37 (1894): 214–218.

10  J. E. Evans and E. W. Maunder, "Experiments as to the Actuality of the 'Canals' Observed on Mars," *Monthly Notices of the Royal Astronomical Society* 63 (1903): 499.

11  Crossley, *Imagining Mars*, 75.

12  Kristina Maria Doyle Lane, *Geographies of Mars: Seeing and Knowing the Red Planet* (University of Chicago Press, 2011).

13  Crossley, *Imagining Mars*, 94.

14  John R. Hammond, *A Preface to H. G. Wells* (New York: Routledge, 2014), 24.

15  David C. Smith, ed., *Correspondence of H. G. Wells, Vol. 1* (London: Pickering and Chatto, 1998), 261.

16  John Taliaferro, *Tarzan Forever: The Life of Edgar Rice Burroughs the Creator of Tarzan* (New York: Scribner, 1999), 15.

17  Crossley, *Imagining Mars*, 237.

18  Markley, *Dying Planet*, 217.

19  Conrad Duncan, "Life on Mars Could Be Found Within Two Years but World Is 'Not Prepared,' Nasa's Chief Scientist Says," *The Guardian*, September 29, 2019, https://www.independent.co.uk/news/science/nasa-mars-life-discovery-space-exomars-rover-chief-scientist-jim-green-a9125076.html.

### CHAPTER TWO—ROCKETEERS

1   James T. Andrews, *Red Cosmos: K. E. Tsiolkovskii, Grandfather of Soviet Rocketry* (College Station: Texas A & M Press, 2009), 16.

2   Tom D. Crouch, *Aiming for the Stars: The Dreamers and Doers of the Space Age* (Washington, D.C.: Smithsonian Press, 1999), 24.

3   Chris Gainor, *To a Distant Day: The Rocket Pioneers* (Lincoln: University of Nebraska Press, 2008), 21.

4   George M. Young, *The Russian Cosmists: The Esoteric Futurism of Nikolai Federov and His Followers* (New York: Oxford University Press, 2012), 51–70.

5   Nikolai F. Fyodorov, "Karazin: Meteorologist or Meteorurge?" in Yvonne Howell, ed. *Red Star Tales: A Century of Russian and Soviet Science Fiction* (Montpelier, VT: Russian Information Services, 2015), 35.

6   Young, *Russian Cosmists*, 45–48.

7   Asif A. Siddiqi, "Imagining the Cosmos: Utopians, Mystics, and the Popular Culture of Spaceflight in Revolutionary Russia," *Osiris* 23 (2005): 266.

8   Young, *Russian Cosmists*, 79.

9   Siddiqi, "Imagining the Cosmos," 266.

10  Young, *Russian Cosmists*, 148.

11  Crouch, *Aiming*, 26.

12  Andrews, *Red Cosmos*, 77.

13  A. A. Blagonravov, ed., *Collected Works of K E. Tsiolkovskiy, Volume II*—"Reactive Flying Machines," 62–63, https://spacemedicineassociation.org/.../Tsiolkovsky%20Oberth%20Goddard.pdf.

14  *Works of K. E. Tsiolkovskiy, Volume II*, 67.

15  *Works of K. E. Tsiolkovskiy, Volume II*, 79.

16  Konstantin Tsiolkovsky, "On the Moon," in Yvonne Howell, ed., *Red Star Tales*, 54.

17  Andrews, *Red Cosmos*, 69.

18  Siddiqi, "Imagining the Cosmos," 268.

19  Robert H. Goddard, "Material for an Autobiography," in Esther C. Goddard and G. Edward Pendray, eds., *The Papers of Robert H. Goddard, Volume One* (New York: McGraw Hill Book Company, 1970), 7–9.

20  The Tom Swift book series began in 1910, but it had many predecessors, including *The Steam Man of the Prairies*, a dime novel first published in 1868, which featured a boy inventor and his marvelous Steam Man.

21  Goddard, "Materials for an Autobiography," 10.

22  Goddard, "Materials for an Autobiography," 13.

23  Goddard, "Materials for an Autobiography," 14.

24  *Papers of Robert H. Goddard*, 95.

25  "R. H. Goddard to Army Chief of Ordnance," August 20, 1917, *Papers of Robert Goddard*, 199.

26  "R. H. Goddard to Secretary, Smithsonian Institution," August 8, 1918, *Papers of Robert Goddard*, 253.

27  Clarence Nichols Hickman, an oral history conducted in 1973 by Julian Tebo and Frank Polkinghorn, IEEE History Center, Hoboken, NJ, https://ethw.org/Oral-History:Clarence_Nichols_Hickman#About_Clarence_Nichols_Hickman.

28  R. H. Goddard "Diary," August 4, 1918, in *Papers of Robert Goddard*, 250.

29  "First Volunteer for Leap to Mars," *New York Times*, February 5, 2020, 1.

30  "Report to Smithsonian Institution Concerning Further Developments of the Rocket Method of Investigating Space," *Papers of Robert Goddard*, 313–30.

31  Cited in William Sims Bainbridge, *The Spaceflight Revolution: A Sociological Study* (New York: John Wiley and Sons, 1976), 32.

32  "Goddard to Secretary, Smithsonian Institution," August 1, 1923, *Papers of Robert Goddard*, 498.

33  Michael J. Neufeld, "Weimar Culture and Futuristic Technology: The Rocketry and Spaceflight Fad in Germany, 1923–1933," *Technology and Culture* 31, no. 4 (1990): 729.

34  Jonathon Keats, "Out of this World: Four Decades before Sputnik, Soviet Avant-Garde Artists Envisioned the Conquest of Space," *Art & Antiques*, February 2016, http://www.artandantiquesmag.com/2016/02/soviet-avant-garde-art/.

35  Siddiqi, "Imagining the Cosmos," 269.

36  Siddiqi, "Imagining the Cosmos," 272.

37  Siddiqi, "Imagining the Cosmos," 277.

38  T. O'Conor Sloane, "Space Travel," *Amazing Stories*, March, 1935, 11.

39  T. O'Conor Sloane, "Discussions," *Amazing Stories*, April 1938, 141.

40  David Clary, *Rocket Man: Robert H. Goddard and the Birth of the Space Age* (New York: Hyperion, 2003), 144–181.

41  Eric Leif Davin, *Pioneers of Wonder* (Amherst, NY: Parthenon Books, 1999), 53.

42  De Witt Douglas Kilgore, *Astrofuturism: Science, Race, and Visions of Utopia in Space* (Philadelphia: University of Pennsylvania Press, 2003), 35.

43  "The ARS Early Years (1930–1944)." American Institute of Aeronautics and Astronautics, https://web.archive.org/web/20150923144229/https://www.aiaa.org/SecondaryTwoColumn.aspx?id=1906.

44  Kilgore, *Astrofuturism*, 31.

## CHAPTER THREE—VON BRAUN

1   Michael J. Neufeld, *Von Braun: Dreamer of Space, Engineer of War* (New York: Knopf, 2008), 24.

2   Bainbridge, *The Spaceflight Revolution*, 49.

3   Wernher von Braun, "Reminiscences of German Rocketry," *Journal of the British Interplanetary Society* vol. 15, no. 3, May–June, 1956, 128.

4   Daniel Lang, "A Romantic Urge," *New Yorker*, April 21, 1951, 83.

5   "Here to Show Us How to Use Mail Rockets," *Brooklyn Daily Eagle,* February 21, 1935, http://afflictor.com/2016/05/25/old-print-article-willy-ley-encourages-america-to-use-mail-rockets-brooklyn-daily-eagle-1935/.

6   Neufeld, *Von Braun*, 55.

7   Michael J. Neufeld, "Creating a Memory of the German Rocket Program for the Cold War," in Stephen J. Dick, ed., *Remembering the Space Age* (NASA, 2008), 74.

8   Von Braun, "Reminiscences," 130.

9   Neufeld, *Von Braun*, 64.

10  Neufeld, *Von Braun*, 144.

11  Neufeld, *Von Braun*, 96–7.

12  Von Braun, "Reminiscences," 143.

13  Bob Ward, *Dr. Space: The Life of Wernher von Braun* (Annapolis, MD: Naval Institute Press, 2005), 40.

14  Neufeld, *Von Braun*, 128.

15  Ward, *Dr. Space*, 50.

16  Ward, *Dr. Space*, 42.

17  G. W. Trichel, Colonel, Ordnance Department, to Colonel Horace Quinn, May 23, 1945, Archives Department, National Air and Space Museum (NASM), Smithsonian Institution, Washington, DC, Richard Porter Papers, Box 5, Folder 15.

18  Seymour Nagan, "Top Secret—Nazis at Work," *New Republic*, August 11, 1947, 24–26.

19  Alfred Africano, "Nazis Vs. U.S. Scientists," *New York Herald Tribune*, February 11, 1947. Library of Congress, Wernher von Braun Papers, Box 53, scrapbooks.

20  These editorials were collected by von Braun. Library of Congress, Wernher von Braun Papers, Box 53, scrapbooks.

21  "Fort Bliss, das Amerikanische Peenemünde" ["Fort Bliss the American Peenemunde"], *Kieler Nachrichten*, April 3, 1948. Wernher von Braun Papers. Box 53, Scrapbooks. Translated for the author by Maxximilian Seijo.

22  A Simpleton, "Jetzt Schlagt's Zwolf," ["Enough is Enough!"] *Berlin am Mittag*,
    July 4, 1947. Wernher von Braun Papers, Box 53. Translated by Maxximilian
    Seijo.

23  "Blitz—Entnazung im Raketentempo!" [Flash—Denazified at Rocket Pace],
    *Leipziger Volkszeitung*, December 12, 1946, Wernher von Braun Papers, Box 53.
    Translated for the author by Maxximilian Seijo.

24  Chesley Bonstell to Wernher von Braun, November 30, 1951, Wernher von
    Braun Papers, Box 42, Folder: Colliers Correspondence and Related Matters.

25  Wernher von Braun to Seth Moseley, March 26, 1952, Wernher von Braun
    Papers, Box 42, Folder: Congress and Space, Colliers.

26  Von Braun to Martin Caider, June 8, 1955, Wernher von Braun Papers, Box 1,
    Folder: 1954 M–Z.

27  "Journey Into Space," *Time*, December 8, 1952, 70.

28  Cited in Ward, *Dr. Space*, 89.

29  Von Braun to John Leonard, October 9, 1953, Wernher von Braun Papers, Box
    1, Folder: 1953 A–H.

30  Wernher von Braun, "We Need a Coordinated Space Program," Fourth
    Congress of the International Astronautic Federation, Zurich, August 1953,
    Wernher von Braun Papers, Box 46, Folder: 1951–55.

31  Wernher von Braun, "Crossing the Last Frontier," *Collier's*, March 22, 1952, 25.

32  Wernher von Braun, "Space Superiority as a Means for Achieving World
    Peace," December 1952. Wernher von Braun Papers, Box 46 "Speeches," Folder:
    1951–1955.

33  Wernher von Braun, "The Importance of Satellite Vehicles in Interplanetary
    Flight," Second International Congress on Astronautics, London, September
    3–8, 1951, Wernher von Braun Papers, Box 46 "Speeches," Folder: 1951–1955.

34  Wernher von Braun, "We Need a Coordinated Space Program."

35  J. P. Telotte, "Disney in Science Fiction Land," *Journal of Popular Film and
    Television* 33, no. 1 (2005), 12–20.

36  "Shooting Script, Prod. 5559 'Rockets & Space,'" Smithsonian Institution,
    NASM Archives Department, Willy Ley Papers, Box 33, Folder 7.

37  Flyer from CIA FOIA files online, CIA-RDP80B01676R003800020019-1,
    https://www.cia.gov/library/readingroom/document/cia-
    rdp80b01676r003800020019-1.

38  Donald Robinson, manuscript of Army article prepared after Explorer launch,
    n.d., circa 1958, Wernher von Braun Papers, Box 46, Folder Jan–April 1958.

39  Wernher von Braun speech, March 18, 1958, Wernher von Braun Papers, Box
    46, Folder: Jan–April 1958.

40  Neufeld, "Creating a Memory," 83.

41  "Remarks by von Braun Before Employees of Marshall Center,"
    July 5, 1961, Wernher von Braun Papers, Box 47, Folder: July–August 1961.

42  Michael Chabon, *Moonglow* (New York: HarperCollins, 2016), 395–401.

43  Von Braun to Richard Porter, August 24, 1969, NASM, SI, Archives
    Department, Richard Porter Papers, Box 3, Folder: Correspondence with von
    Braun.

CHAPTER FOUR—MODERN CONQUERORS OF MARS

1   Arthur C. Clarke, "What Will We Do with the Moon?" *Popular Science*, April, 1952, 166.

2   Otto O. Binder, "Mars Colony," *Space World* 1, no. 8, (1961): 31.

3   *Mars Terraforming not Possible Using Present-Day Technology*, NASA, July 20, 2018, https://www.nasa.gov/press-release/goddard/2018/mars-terraforming.

4   Jim Zimbelman, of the Smithsonian Institution, calculated that the feet of Wilt Chamberlain, who had the highest vertical jump in basketball, could have cleared the basketball rim by about five inches on Mars. Zimbelman, email to author, March 23, 2020.

5   Markley, *Dying Planet*, 221.

6   Carl Sagan, "Planetary Engineering," *Icarus* vol. 20, issue 4, December 1973, 513–14.

7   M. M. Averner and Macelroy, "On the Habitability of Mars: An Approach to Planetary Ecosynthesis," Washington, DC: NASA technical report, January 1, 1976.

8   Christopher McKay, et al., "Making Mars Habitable," *Nature*, August 8, 1991, 489–496.

9   Wernher von Braun, "The Importance of Satellite Vehicles in Interplanetary Travel," Second International Congress on Astronautics, September 3–8, 1951. Wernher von Braun Papers.

10  Robert Zubrin, *The Case for Mars: The Plan to Settle the Red Planet and Why We Must* (New York: Free Press, 1996, 2011), 267.

11  John Wenz, "There's not enough CO2 to Terraform Mars," *Astronomy*, July 31, 2018, http://www.astronomy.com/news/2018/07/terraforming-mars-is-a-no-go.

12  Andrew Coates, "Sorry Elon . . . ," *Independent*, August 6, 2018, https://www.independent.co.uk/voices/elon-musk-mars-colony-space-tesla-terraforming-a8479156.html.

13  Adam Rogers, "Sorry Nerds: Terraforming Might Not Work on Mars," *Wired.com*, July 30, 2018, https://www.wired.com/story/co2-terraforming-mars/.

14  Leah Crane, "Terraforming Mars might be impossible due to a lack of carbon dioxide," *New Scientist*, July 30, 2018, https://www.newscientist.com/article/2175414-terraforming-mars-might-be-impossible-due-to-a-lack-of-carbon-dioxide/.

15  Ashlee Vance, *Elon Musk: Tesla, SpaceX, and the Quest for a Fantastic Future* (New York: HarperCollins, 2015), 24.

16  Vance, 37.

17  With no specific budget lines, NASA has always been quite vague about the schedule for manned missions to Mars. As of 2018, there still was no timetable but the idea that by the 2040s the manned mission would take place.

18  Vance, 4.

19  Musk receiving 2012 Mars Pioneer Award at Mars Society Conference, Accessed August 1, 2018, https://www.youtube.com/watch?v=PK0kTcJFnVk.

20  A few months after the conference, Tesla reported a profit for its third quarter in 2018, suggesting its economic woes were on the wane and the Model-3 was selling well.

21  Palmer Luckey, "Because He Is a Superhero," from "In Defense of Elon Musk," *Popular Mechanics*, October 16, 2018, https://www.popularmechanics.com/space/rockets/a23508636/defense-of-elon-musk/.

22  Zubrin interview in Poland, September 17, 2018, https://www.rmf24.pl/nauka/news-pewnegodnia-mars-bedzie-bardzo-ekscytujacym-miejscem-do-zyc,nId,2632545.

23  Wernher von Braun, *Project Mars: A Technical Tale* (Burlington, Canada: Apogee Books, 2004 [1949]), 177.

24  "Mission," Mars One, https://www.mars-one.com/mission (Accessed August 15, 2018).

25  Jonathan O'Callaghan, "It's Not Going to Be Big Brother on Mars," *Daily Mail*, February 20, 2015, https://www.dailymail.co.uk/sciencetech/article-2961743/It-s-not-going-Big-Brother-Mars-exciting-story-time-Mars-One-founder-remains-convinced-going-red-planet.html.

26  Jennifer Chu, "Mars One (And Done?)" October 14, 2014, http://news.mit.edu/2014/technical-feasibility-mars-one-1014.

27  Anna Holligan, "Can the Dutch Do Reality TV in Space?" BBC News, June 20, 2012, https://www.bbc.com/news/world-europe-18506033.

28  Andrew Smith, "Can Mars One Colonize the Red Planet?" *The Guardian*, May 30, 2015, https://www.theguardian.com/science/2015/may/30/can-mars-one-colonise-red-planet.

29  Gwendolyn Smith, "Mars One Mission: 'My Boyfriend Is Cool with Me Going to Mars on a One-Way Trip,'" *The Telegraph*, November 13, 2015, https://www.telegraph.co.uk/women/life/mars-one-mission-my-boyfriend-is-cool-with-me-going-to-mars-on-a-one-way-trip/.

30  Interview with Dianne McGrath, on podcast, "Future Martians: Interviews with the Mars 100, Episode 10," https://plus.google.com/+Mars-one/posts/chJE9pLQ7dn (Accessed September 10, 2018).

31  Interview with Dianne McGrath.

32  Karl Tate, "Space Radiation Threat to Astronauts Explained," *Space.com*, May 30, 2018, https://www.space.com/21353-space-radiation-mars-mission-threat.html.

33  Kim Stanley Robinson, email to author, March 13, 2019.

34  Robert Zubrin, email to author, March 16, 2019.

35  J. R. Skok, phone interview with author, October 2, 2018.

## CHAPTER FIVE—SPACE COLONIES

1  Patrick McCray, *Visioneers: How a Group of Elite Scientists Pursued Space Colonies, Nanotechnologies, and a Limitless Future* (Princeton University Press, 2017).

2  Tasha O'Neill, phone interview with author, September 25, 2019.

3  Al Reinart, "Gerry's World," *Air & Space*, April/May 1989, 30.

4  Gerard O'Neill to Walker Bleakney, July 20, 1967, Archives Department, NASM, SI, Washington, DC, Gerard K. O'Neill Collection Box 5, Folder 9.

5  George Pimentel to Gerard K. O'Neill, July 9, 1967, Archives Department, SASM, SI, Gerard K. O'Neill Collection, Box 5, Folder 9.

6   Gerard O'Neill, *The High Frontier* (New York: William Morrow and Company, 1977), 236.

7   "Public Trust in Government: 1958–2019," Pew Research Center, April 11, 2019, https://www.people-press.org/2019/04/11/public-trust-in-government-1958-2019/.

8   Cary Fun and Brian Kennedy, "Public Confidence in scientists has remained stable for Decades," Pew Research Center, March 22, 2019, https://www.pewresearch.org/fact-tank/2019/03/22/public-confidence-in-scientists-has-remained-stable-for-decades/.

9   Donnela H. Meadows, Dennis L. Meadows, Jorgen Randers, William W. Behrens III, *Limits to Growth* (New York: Universe Books, 1972), 25.

10  Meadows, et al., *Limits to Growth*, 127–8.

11  Herb Bernstein, email to author, August 21, 2019.

12  *High Frontier*, 248.

13  T. A. Heppenheimer, *Colonies in Space* (Harrisburg, PA: Stackpole Books, 1976), 34–49.

14  Tasha O'Neill, phone interview, September 25, 2019.

15  O'Neill, *High Frontier*, 43.

16  Stewart Brand, ed., *Space Colonies* (New York: Penguin Books, 1977), 23.

17  Arthur C. Clarke, "Rocket to the Renaissance," in *Greetings, Carbon-Based Bipeds! Collected Essays, 1934–1988* (New York: St. Martin's Press, 1999), 214.

18  Brian O'Leary, *The Fertile Stars* (New York: Everest House, 1981), 93–94.

19  O'Neill, *High Frontier*, 170.

20  George Koopman, "We're Going," *L5 News*, September 1977, 2–3.

21  Robert Crumb, email to author, September 9, 2019.

22  "Inner Space and Outer Space: Carl Sagan's Letters to Timothy Leary," Timothy Leary Archives.org, http://www.timothylearyarchives.org/carl-sagans-letters-to-timothy-leary-1974/.

23  *L5-News*, October 1976, 11.

24  *L5-News*, December 1976.

25  Brand, *Space Colonies*, 72.

26  De Witt Douglas Kilgore, *Astrofuturism: Science, Race, and Visions of Utopia in Space* (Philadelphia: UPenn Press, 2003), 169.

27  Kilgore, *Astrofuturism*, 52.

28  McCray, *Visioneers*, 107.

29  Brand, *Space Colonies*, 45

30  Brand, *Space Colonies*, 44.

31  Brand, *Space Colonies*, 40.

32  Brand, *Space Colonies*, 44.

33  "The Gee Whiz Society," Space Studies Institute, December 18, 2019, http://ssi.org/gee-whiz-society/.

34  Gerard K. O'Neill Collection, SASM, SI, Box 55, Folder 16.

35  O'Neill's lecture notes, "2081," n.d., likely presented in 1981, NASM, SI, Gerard K. O'Neill Collection, Box 53, Folder 1.

36  Tasha O'Neill, phone interview, September 25, 2019.

37  O'Neill to Robert Hofstadter, August 25, 1983, NASM, SI, Gerard K. O'Neill Collection, Box 51, folder 9.

38  T. A. Heppenheimer, "Astrolling the Astroturf," *L-5 News*, January 1976.

39  *L-5 News*, October 1977, 8.

40  Bruce Friedman, "L-5 and the Jewish Community," *L-5 News*, October 1977, 14.

41  Jackie Knowles, "Paradise in Heavens Sought by Space Club," *Pasadena Star-News*, December 11, 1977, 1.

42  Daren Nigsarian, email to author, August 19, 2019.

43  Taylor Dark, email to author, July 20, 2018.

44  "Ten Years of Perspective on Space Colonization," in Charles H. Holbrow, Allen M. Russell, Gordon F. Sutton, eds., *Space Colonization: Technology and the Liberal Arts* (New York: American Institute of Physics, 1986), 142.

45  "Ten Years of Perspective on Space Colonization," 143.

46  "Ten Years of Perspective on Space Colonization," 147; 150–52.

47  "Ten Years of Perspective on Space Colonization," 153.

48  Carolyn Meinel, email to author, September 26, 2019.

49  Alan Boyle, "Jeff Bezos: 'We will have to leave this planet . . . and it's going to make this planet better,'" *Geekwire*, May 29, 2019, https://www.geekwire.com/2018/jeff-bezos-isdc-space-vision/.

50  Mark Hopkins, "A $178 Million Infusion for Space Solar Power," *ad Astra* 1 (2019): 13.

51  Taylor Dark, email to author, April 9, 2019.

52  Carolyn Meinel, email to author, September 26, 2019.

53  "Ten Years of Perspective on Space Colonization," 145.

54  "Ten Years of Perspective on Space Colonization," 146.

CHAPTER SIX—BIOSPHERE 2

1   John Allen, *Me and the Biospheres: A Memoir by the Inventor of Biosphere 2* (Santa Fe: Synergetic Press, 2009), vii.

2   Allen, *Me and the Biospheres*.

3   Allen, *Me and the Biospheres*, 1.

4   Allen, *Me and the Biospheres*, 9.

5   Rebecca Reider, *Dreaming the Biosphere: The Theater of All Possibilities* (University of New Mexico Press, 2009), 32.

6   Jane Poynter, *The Human Experiment: Two Years and Twenty Minutes Inside Biosphere 2* (New York: Thunder's Mouth Press, 2006), 18, 21.

7   Poynter, *The Human Experiment*, 39.

8   Reider, *Dreaming the Biosphere*, 23.

9   Reider, *Dreaming the Biosphere*, 30–31.

10  Dan Burton and David Grandy, *Magic, Mystery, and Science: The Occult in Western Civilization* (Bloomington: Indiana University Press, 2004), 318.

11  See Reider, *Dreaming the Biosphere*, 115–6.

12  Laurence Veysey, *The Communal Experience: Anarchist and Mystical Counter-Cultures in America* (New York: Harper and Row, 1973), 279.

13  Allen, *Me and the Biospheres*, 46.
14  Reider, *Dreaming the Biosphere*, 271.
15  Poynter, *The Human Experiment*, 19–20.
16  William J. Broad, "As Biosphere Is Sealed, Its Patron Reflects on Life," *New York Times*, September 24, 1991.
17  Joel Achenbach, "Bogus New World," *Washington Post*, January 8, 1992.
18  Confirmed by the Smithsonian Institution's Jim Zimbelman and other planetary science staff via email, September 18, 2019.
19  John Allen and Mark Nelson, *Space Biospheres* (Oracle, AZ: Synergetic Press, 1989), 1.
20  Allen and Nelson, *Space Biospheres*, 82.
21  Young, *The Russian Cosmists*, 157.
22  Allen and Nelson, *Space Biospheres*, 53.
23  Allen, *Me and the Biospheres*, 93.
24  I. I. Gitelson, I. A. Terskov, et al., "Long Term Experiments on Man's Stay in Biological Life-Support System," in *Controlled Ecological Life Support System: Natural and Artificial Ecosystems*, NASA Conference Publication 10040, 63.
25  D. Obenhuber and C. Folsome, "Carbon Recycling in Materially Closed Ecological Life Support Systems," *Bio Systems* no. 21, 1988, 165–73, https://www.researchgate.net/scientific-contributions/4883299_Clair_E_Folsomee.
26  Lynn Margulis and Dorion Sagan, *What Is Life?* (Berkeley: University of California Press, 2000), 235.
27  Allen, *Me and the Biospheres*, 180.
28  Jeremy Walker and Celine Granjou, "MELiSSA the Minimal Biosphere: Human Life, Waste and Refuge in Deep Space," *Futures*, May 2017, https://www.researchgate.net/publication/311825298_MELiSSA_the_minimal_biosphere_Human_life_waste_and_refuge_in_deep_space.
29  Reider, *Dreaming the Biosphere*, 71.
30  Allen, *Me and the Biospheres*, 127.
31  Mark Nelson, *Pushing Our Limits: Insights from Biosphere 2* (Tuscon: University of Arizona Press, 2018), xiv.
32  Reider, *Dreaming the Biosphere*, 90.
33  Veysey, *The Communal Experience*, 293.
34  Veysey, *The Communal Experience*, 294.
35  Marc Cooper, "Take This Terrarium and Shove It," *Village Voice*, April 2, 1991, 25.
36  Cooper, "Take This Terrarium and Shove It," 29.
37  Seth Mydans, "8 Sealed in a World beneath Glass for Two Years," *New York Times*, September 27, 1991.
38  Carlyle C. Douglas, "A Voyage of Discovery that Doesn't Move," *New York Times*, September 20, 1991, E7.
39  Reider, *Dreaming the Biosphere*, 138.
40  Poynter, *The Human Experiment*, 130.
41  "8 Scientists Locked into Biosphere for 2-Year Test," *Los Angeles Times*, September 27, 1991, A35.

42  Poynter, *The Human Experiment*, 5.

43  Achenbach, "Bogus New World."

44  Poynter, *The Human Experiment*, 24.

45  Reider, *Dreaming the Biosphere*, 189.

46  E. K. Erik Gunderson, *Mental Health Problems in Antarctica, Report No. 67-14*, Bureau of Medicine and Surgery Department of the Navy, Reprinted from the Archives of Mental Health, October 1968, vol. 17.

47  Robert Reinhold, "Strife and Despair at South Pole Illuminate Psychology of Isolation," *New York Times*, January 12, 1982, https://www.nytimes.com/1982/01/12/science/strife-and-despair-at-south-pole-illuminate-psychology-of-isolation.html.

48  Reider, *Dreaming the Biosphere*, 190.

49  Seth Mydans, "8 Bid Farewell to 'Future': Stale Air, Roaches and Ants," *New York Times*, September 27, 1993, A1, 13.

50  Tony Burgess, "Refuge Notebook: Species Packing into Biosphere 2," *Peninsula Clarion*, November 18, 2018, https://www.peninsulaclarion.com/sports/refuge-notebook-species-packing-into-biosphere-2/.

51  George de Lama, "Biosphere Has Coming out Party," *Chicago Tribune*, September 27, 1993, D1, 3.

52  Robert Lee Hotz, "Biosphere Crew to End a Two Year Stay in a World of Their Own," *Los Angeles Times*, September 25, 1993, A1, 22–3.

53  George de Lama, "Biosphere Has Coming out Party," 8.

54  Reider, *Dreaming the Biosphere*, 210.

55  "Some Very This Worldly Trouble," *The Chicago Tribune*, 6 April 1994, N14.

56  Dave Mosher, "Life in a Bubble: How We Can Fight Hunger, Loneliness, and Radiation on Mars," *Business Insider*, April 12, 2018, https://www.businessinsider.com/mars-colonization-human-survival-radiation-food-biosphere-2018-4.

## CHAPTER SEVEN—MAKING SPACE FUN AGAIN

1  Rachel Arthur, "Fashion's Space Race: Why the Spacesuit Is a Huge Future Branding Opportunity for Designers," *Forbes*, August 27, 2017, https://www.forbes.com/sites/rachelarthur/2017/08/27/fashions-space-race/#557e88417dff.

2  John Spencer, phone interview with author, May 6, 2019.

3  "Las Vegas Visitor Profile Survey," Las Vegas Convention and Visitors Authority, 2018, 29, https://www.lvcva.com/stats-and-facts/visitor-profiles/.

4  John Spencer, Mensa presentation, 2016, https://www.youtube.com/watch?v=OCxvANETd6w.

5  Helene Cooper, R. Blumenthal, and L. Kean, "Glowing Auras and Black Money: Then Pentagon's Mysterious UFO Program," *New York Times*, December 16, 2017, https://www.nytimes.com/2017/12/16/us/politics/pentagon-program-ufo-harry-reid.html.

6  Lara Logan, "Bigelow Aerospace Founder Says Commercial World Will Lead in Space," *Sixty Minutes*, May 28, 2017, https://www.cbsnews.com/news/bigelow-aerospace-founder-says-commercial-world-will-lead-in-space/.

7   "FAQ on Space Tourism and Exploration," Spaceport Associates website, http://www.spaceportassociates.com/faq.html.

8   David B. Yaden, Jonathan Iwry, et. al, "The Overview Effect: Awe and Self-Transcendent Experience in Space Flight," *Psychology of Consciousness: Theory, Research, and Practice* vol. 3, no.1, 2016, 1.

9   Yaden, Iwry, et. al, "The Overview Effect, 6.

10  "Declaration of Vision and Principles," The Overview Institute, https://overviewinstitute.org/about-us/declaration-of-vision-and-principles/.

11  Maureen O'Hare, "Look inside the first luxury space hotel," CNN, January 21, 2019, https://www.cnn.com/travel/article/aurora-station-luxury-space-hotel/index.html.

12  Samantha Masunaga, "Now Boarding Space Tourists—for a Price," *Los Angeles Times*, June 8, 2019, C1, 5.

13  "Space Exploration: Can Sex, Drugs and Rock'n'Roll Replace Flag Planting?" *San Jose Mercury News*, July 24, 2009, http://www.parabolicarc.com/2009/07/24/space-exploration-sex-drugs-rocknroll-replace-flag-planting/#more-7253.

14  David Kushner, "Going to Space? First Stop: Eight Months of Grueling Training in Russia's Star City," *Wired*, August 18, 2008, https://www.wired.com/2008/08/ff-starcity/.

15  Alex Layendecker, phone interview with author, May 25, 2019.

16  Alexander Layendecker and S. Pandya, "Logistics of Reproduction in Space," in E. Seedhouse and D. J. Shayler, eds., *Handbook of Life Support Systems for Spacecraft and Extraterrestrial Habitats* (Basel: Springer Nature, 2019), https://doi.org/10.1007/978-3-319-09575-2_211-1.

17  Layendecker, *Sex In Outer Space and the Advent of Astrosexology* (PhD dissertation), Institute for Advanced Study of Human Sexuality, May 1, 2016, 80.

18  Rick Hampson, "NASA Told to Make Room for Sex in Space," AP News, June 25, 1985, https://www.apnews.com/0b6a54cedbf96659356945815d3d431e.

19  Layendecker, "Sex in Outer Space," 108.

20  Layendecker, "Sex in Outer Space," 109.

21  Thilini P. Schlesinger, et al., "International Space Station Crew Quarters On-Orbit Performance and Sustaining," NASA, 2013, 2, https://ntrs.nasa.gov/archive/nasa/casi.ntrs.nasa.gov/20130011142.pdf.

22  Layendecker and Pandya, "Logistics," 4.

23  Layendecker, "Sex in Outer Space," 108.

24  Bonta also invented a "Jet shoe" that shifts from a "flat" to a "pump." She told one reporter, "A pump is dangerous in zero gravity . . . But when you land, you want attractive footwear." Jennifer Adkins, "Sex and the Stratosphere," *Inventors' Digest,* May 2009, https://web.archive.org/web/20130409125140/http://www.inventorsdigest.com/archives/591.

25  Adkins, "Sex and the Stratosphere," 5.

26  "Personal Statement: Kees Mulder," Spacelife Origin, https://spacelifeorigin.com/en (accessed July 22, 2019).

27  Tod Martens, "New Force Hits Town," *Los Angeles Times*, June 2, 2019, E8.

28  "Astronaut Training Experience," Kennedy Space Center Visitor Complex, https://www.kennedyspacecenter.com/explore-attractions/all-attractions/astronaut-training-experience.

29  "Mars Base 1," Kennedy Space Center Visitor Complex, https://www.kennedyspacecenter.com/explore-attractions/all-attractions/mars-base-1.

30  Sandra Häuplik-Meusburger, Kim Binsted, et al., "Habitability Studies and Full Scale Simulation Research: Preliminary Themes following HISEAS mission IV," 47th International Conference on Environmental Systems (conference paper), July 2017, 2.

31  Lisa Spence, phone interview with author, June 21, 2019.

32  Lindsay Larson, Leslie DeChurch, Suzanne T. Bell, Noshir Contractor, "Team performance in space crews: Houston, we have a teamwork problem," *Acta Astronautica*, May 2019, https://www.researchgate.net/publication/332854348.

33  Marina Koren, "The Asteroid Misson that Never Leaves Earth," *The Atlantic*, December 22, 2017, https://www.theatlantic.com/science/archive/2017/12/wow-its-cramped-in-here/549032/.

34  Lisa M. P. Munoz, "Brains in Space: The Important Role of Cognitive Neuroscience in Deep-Space Missions," *Journal of Cognitive Neuroscience*, December 21, 2017, https://www.cogneurosociety.org/brains-in-space-cognitive-neuroscience/.

35  Munoz, "Brains in Space."

36  Christian Davenport, "How Much Does a Ticket to Space Cost? Meet the People Ready to Fly," *Washington Post*, October 2, 2019, https://www.washingtonpost.com/technology/2019/10/02/how-much-does-ticket-space-cost-meet-people-ready-fly/.

37  Kenneth Chang, "Virgin Galactic Unveils Jumpsuit for Space Tourists," *New York Times*, October 16, 2019, https://www.nytimes.com/2019/10/16/science/virgin-galactic-spacesuit.html?action=click&module=Editors%20Picks&pgtype=Homepage.

38  Oliver Morton, *The Moon: A History for the Future* (New York: Public Affairs/Hachette Books, 2019), 215–220.

## CHAPTER EIGHT—THE MOON

1  Arlin Crotts, *The New Moon: Water, Exploration, and Future Habitation* (New York: Cambridge University Press, 2014), 2.

2  James Joyce, *Ulysses* (Mineola, NY: Dover Books, 2018), 679.

3  "The Myth of Moon Phases and Menstruation," *Clue*, July 14, 2016, https://helloclue.com/articles/cycle-a-z/myth-moon-phases-menstruation.

4  C. P. Thakur and Dilip Sharma, "Full Moon and Crime," *British Medical Journal* (Clinical Research Edition) 289, no. 6460 (1984): 1789–791, http://www.jstor.org/stable/29517732.

5  Kepler, *Somnium*. Frosty Drew Observatory, https://frostydrew.org/papers.dc/papers/paper-somnium/.

6  William D. Moore, "Riding the Goat: Secrecy, Masculinity, and Fraternal High

Jinks in the United States, 1845–1930," *Winterthur Portfolio* vol. 41, no. 2/3 (2007), 164.

7   Moore, "Riding the Goat," 395.

8   Memo, JFK to Lyndon Johnson, April 20, 1961, https://history.nasa.gov/ Apollomon/docs.htm.

9   Memo, von Braun to Johnson, April 29, 1961, https://history.nasa.gov/ Apollomon/docs.htm.

10  Memo, Lyndon Johnson to JFK, April 28, 1961, https://history.nasa.gov/ Apollomon/docs.htm.

11  John F. Kennedy, "Excerpts from 'Urgent National Needs Speech to a Joint Session of Congress, 25 May 1961,'" https://history.nasa.gov/Apollomon/docs. htm.

12  Norman Mailer, *Of a Fire on the Moon* (Boston: Little, Brown and Company, 1969), 344–45.

13  Mailer, 372–381.

14  Frederick Ordway III, et al., "Project Horizon: An Early Study of a Lunar Outpost," *Acta Astronautica* 17, no. 10 (1988), 1109.

15  "Project Horizon, Volume One: Summary and Supporting Considerations," U.S. Army, 1959, 15.

16  "Project Horizon," 81.

17  Ordway, "Project Horizon," 1118.

18  "Propose Moon Base," *Science News Letter,* May 30, 1959, Archives Department, NASM, SI, Technical Files, Space Colonies, Folder 649150-01.

19  Arthur J. Snider, "Easy Life for Moon Colonists," *Philadelphia Inquirer*, October 12, 1960.

20  Thomas R. Phillips, "Value of a Moon Base for Military Operations," *St. Louis Post-Dispatch*, September 27, 1959, 29.

21  Von Braun to Otto Binder (editor of *Space World*), September 25, 1961, Library of Congress, Wernher von Braun Collection, Box 5, Folder: 1961B.

22  "Red Says Moon Won't Be Armed," *Washington Star*, October 5, 1959, 6.

23  Asif Siddiqi, email to author, January 20, 2020.

24  Oleg Yegorov, "Why Did Soviets (And Americans) Want to Nuke the Moon?" Russia Beyond, August 29, 2019, https://www.rbth.com/history/330892-ussr-usa-nuclear-bomb-moon.

25  A. Perry, "Moon City Planning—Hobby of Red Astronautic Civil-Defense Chief," n.d., Archives Department, NASM, SI technical folders, "Space, Colonies, Lunar, Russia," 649240-01.

26  Gerard J. De Groot, *Dark Side of the Moon: The Magnificent Madness of the American Lunar Quest* (New York University Press, 2006), 234.

27  Bryan Green, "While NASA Was Landing on the Moon, Many African-Americans Sought Economic Justice Instead," *Smithsonian*, July 11, 2019, https://www.smithsonianmag.com/history/nasa-landing-moon-many-african-americans-sought-economic-justice-instead-180972622/.

28  Steven Morris, "How Blacks View Mankind's 'Giant Step,'" *Ebony*, September 1970, 33.

29  William Sims Bainbridge, "The Impact of Space Exploration on Public Opinions, Attitudes, and Beliefs," in Steven J. Dick, ed. *Historical Studies in the Societal Impact of Spaceflight* (Washington, DC: NASA, 2015), 15.

30  Crotts, *The New Moon*, 340.

31  Peter Kokh, "Of Milestones and Goals," *Moon Miners' Manifesto*, no. 23, March 1989.

32  Crotts, *New Moon*, 348.

33  Ian McDonald, *Luna: New Moon* (New York: TOR, 2015), 19.

34  Peter Kokh, "Prinzton, a Rille-Bottom Settlement for Three Thousand People," *Moon Miner's Manifesto*, no. 26, July 1989.

35  L. A. Taylor and T. T. Meek, "Microwave Sintering of Lunar Soil: Properties, Theory, and Practice," *Journal of Aerospace Engineering* vol 18, issue 3, July 2005.

36  Peter Kokh, telephone interview with author, August 30, 2019.

37  Andy Weir, email to author, August 19, 2019.

38  Alan Boyle, "NASA Could Go Back to the Future and Revive Moon Base Plans from a Decade Ago," *Geekwire*, October 6, 2017, https://www.geekwire.com/2017/nasa-go-back-future-revive-moon-base-plans-decade-ago/.

39  Deborah Netburn, "Soon, a Female Moonwalk," *Los Angeles Times*, July 20, 2019, A1.

40  As of this writing, NASA and the administration were pushing for a Moon landing by 2024, while Congress preferred 2028.

41  Netburn, "Soon, a Female Moonwalk," A8.

42  Fiona MacDonald, "NASA Scientists Say We Could Colonise the Moon by 2022 . . . For Just $10 Billion," *Science Alert*, March 22, 2016, https://www.sciencealert.com/nasa-scientists-say-we-could-colonise-the-moon-by-2022-for-just-10-billion.

43  Eric Berger, "Buzz Aldrin Is Looking Forward, Not Back—and He Has a Plan to Bring NASA Along," July 9, 2019, *Arstechnica*, https://arstechnica.com/science/2019/07/buzz-aldrin-is-looking-forward-not-back-and-he-has-a-plan-to-bring-nasa-along/.

44  Jeff Faust, "NASA's Lunar Space Station Is a Great/Terrible Idea," *Spectrum: IEEE Journal*, July 8, 2019, https://spectrum.ieee.org/aerospace/space-flight/nasas-lunar-space-station-is-a-greatterrible-idea?fbclid=IwAR1EF89SLPI8vZz3X6-lV6Lx_wVGGdd2QgOMUtX5DgtE26CTzZ0plzEjYqo.

45  Andy Weir, email to author, August 19, 2019.

46  Dale Skran, email to author, August 2, 2019.

47  Jeff Foust, "NASA Takes Gateway Off the Critical Path for 2024 Lunar Return," March 13, 2020, *SpaceNews*, https://spacenews.com/nasa-takes-gateway-off-the-critical-path-for-2024-lunar-return/.

48  Robert Zubrin, email to author, March 19, 2020. This book went to press before further NASA announcements.

49  Oliver Morton, *The Moon: A History of the Future* (New York: Public Affairs, 2019), 272.

50  Crotts, *New Moon*, 364.

51  Jan Woerner, "A Vision for Global Cooperation and Space 4.0," European Space Agency Press Release, 2016.

52  Colin Koop and Daniel Innocente, phone interview with author, November 19, 2019.

## CHAPTER NINE—GOING INTERSTELLAR

1   "The Ultimate Migration," British Interplanetary Society, https://www.bis-space.com/2012/03/23/4110/the-ultimate-migration.

2   Louis Friedman, *Human Spaceflight: From Mars to the Stars* (Tucson: The University of Arizona Press, 2015), 51.

3   J. D. Bernal, *The World, the Flesh, and the Devil: An Enquiry into the Future of the Three Enemies of the Rational Soul, 2nd edition,* (Bloomington: Indiana University Press, 1969 [1929]),18–24.

4   Bernal, *The World, the Flesh, and the Devil*, 28.

5   Bernal, *The World, the Flesh, and the Devil*, 29.

6   Bernal, *The World, the Flesh, and the Devil* 42–46.

7   E. Mallove and G. Matloff, *The Starflight Handbook: A Pioneer's Guide to Interstellar Travel* (New York: John Wiley & Sons, 1989), 4.

8   Mallove and Matloff, *The Starflight Handbook*, 24–25.

9   Mallove and Matloff, *The Starflight Handbook*, 173.

10  Nikos Prantzos, *Our Cosmic Future: Humanity's Fate in the Universe* (Cambridge University Press, 2000), 131.

11  Jatila van der Veen, email to author, December 6, 2019.

12  James Benford, "Sailships vs. Fusion Rockets: A Contrarian View," *Journal of the British Interplanetary Society* 70, (2017): 175.

13  Paul Gilster, *Centauri Dreams: Imagining and Planning Interstellar Travel* (New York: Copernicus Books, 2004), 4.

14  Freeman Dyson, phone interview with author, October 3, 2019.

15  Nikos Prantzos, *Our Cosmic Future: Humanity's Fate in the Universe* (Cambridge University Press, 2000), 105.

16  James Benford, "Sailships," 175.

17  Mallove and Matloff, *The Starflight Handbook*, 195–7.

18  Michio Kaku, *The Future of Humanity: Terraforming Mars, Interstellar Travel, Immortality and Our Destiny Beyond Earth* (New York: Doubleday, 2018), 14.

19  Kaku, *The Future of Humanity*, 268.

20  Miguel Alcubierre, "The Warp Drive: Hyper-Fast Travel within General Relativity," *Classical and Quantum Gravity* 11, no. 5 (1994): 73–77.

21  See Kaku, *The Future of Humanity*, 165, and Gilster, *Centauri Dreams*, 4.

22  Adam Frank, "Are We About to Send Spacechips to the Stars?" NPR 13.7 "Cosmos and Culture," May 10, 2016, https://www.npr.org/sections/13.7/2016/05/10/477448335/are-we-about-to-send-spacechips-to-the-stars.

23  Joel Davis, "With Antimatter to the Stars," *New Scientist*, June 24, 1989, 67.

24  Geoffrey A. Landis, "Advanced Solar and Laser-Pushed Lightsail Concepts," NASA Institute for Advanced Concepts, May 31, 1999, 3.

25  James Benford, email to author, February, 13, 2020.

26  James Benford, "Starship Sails Propelled by Cost-Optimized Directed Energy," *Journal of the British Interplanetary Society* 66, (2013): 85.

27  Philip Lubin, March 14, 2019 lecture at Santa Barbara City College.

28  Philip Lubin, March 14, 2019 lecture.

29  Philip Lubin, interview with author, March 3, 2019.

30  Philip Lubin, interview with author, March 3, 2019.

31  Benford, "Sailships vs. Starships," 177.

32  Freeman Dyson, interview with author, October 3, 2019.

33  Rachel Armstrong, *Star Ark: A Living, Self-Sustaining Spaceship* (Chichester, UK: Springer-Praxis, 2017), i.

34  Armstrong, *Star Ark*, 141.

35  Armstrong, *Star Ark*, 91.

36  Armstrong, *Star Ark*, 28.

37  Armstrong, *Star Ark*, 30.

38  Armstrong, *Star Ark*, 33.

39  Freeman Dyson, interview with author, October 3, 2019.

40  Martin Rees, "To the Ends of the Universe," in *Starship Century*, 435.

41  Friedman, *Human Spaceflight*, 6.

42  Friedman, *Human Spaceflight*, 5.

43  Meghan Neal, "Our Best Bet for Colonizing Space May Be Printing Humans on Other Planets," *Motherboard*, May 29, 2014, https://motherboard.vice.com/en_us/article/ypwwpy/our-best-bet-for-colonizing-space-may-be-printing-humans-on-other-planets.

44  Kim Stanley Robinson, *Aurora* (New York: Orbit, 2015), 178.

45  Robinson, *Aurora*, 179.

46  Robinson, *Aurora*, 418.

47  Kim Stanley Robinson, email interview with the author, March 13, 2019.

48  Geoff Ryman, et al., "The Mundane Manifesto," 2004, https://sfgenics.wordpress.com/2013/07/04/geoff-ryman-et-al-the-mundane-manifesto/.

## CHAPTER TEN—THE METAPHYSICAL LURE OF DEEP SPACE

1  Excerpt from John F. Kennedy acceptance speech of democratic presidential nomination, July 15, 1960, https://www.jfklibrary.org/asset-viewer/archives/JFKSEN/0910/JFKSEN-0910-015.

2  O'Neill, *High Frontier*, 205.

3  Zubrin, *The Case for Mars*, 304.

4  Kaku, *The Future of Humanity*, 35.

5  Krafft A. Ehricke and Elizabeth A. Miller, "From Closed to Open World: The Extraterrestrial Imperative," *EIR*, August 26, 2011, 38.

6  Matthew H. Nitecki, "Discerning the Criteria for Concepts of Progress," in Matthew H. Nitecki, ed., *Evolutionary Progress* (University of Chicago Press, 1988), 6.

7   From a 1984 speech by Ehricke, cited in Joseph A. Angelo, *Space Technology* (Westport, CT: Greenwood Press, 2003), 282.

8   Joseph Kirby, "Toward an Ecological and Cosmonautical Philosophy," *Journal of Evolution & Technology* 23, no. 1 (2013): 5.

9   Michael Ruse, "Molecules to Men: Evolutionary Biology and Thoughts of Progress," in Matthew H. Nitecki, ed., *Evolutionary Progress* (University of Chicago Press, 1988), 97.

10  Stephen J. Gould, "On Replacing the Idea of Progress with an Operational Notion of Directionality," in Matthew H. Nitecki, ed., *Evolutionary Progress*, 319.

11  Ehricke and Miller, "From Closed to Open World," 40.

12  Lee Billings, "It's Never Aliens—Until it Is," *Scientific American*, January 9, 2018, https://www.scientificamerican.com/article/its-never-aliens-until-it-is/.

13  William A. Gale and Gregg Edwards, "Models of Long Range Growth," in William A. Gale, ed., *Life in the Universe: The Ultimate Limits of Growth* (Boulder, CO: Westview Press, 1979), 71.

14  Gale and Edwards, "Models of Long Range Growth," 93.

15  Arthur C. Clarke, "The Challenge of the Spaceship," in *Greetings Carbon-Base Bipeds! Collected Essays 1934–1998* (New York: St. Martin's Press, 2000), 34.

16  Teilhard de Chardin, *Activation of Energy* (New York: Harcourt Brace Jovanovich, 1970), 273.

17  Charles Collins, "Actor Finds Document Teilhard Forced to Sign Affirming Church Teaching," *The Crux*, February 23, 2018, https://cruxnow.com/church-in-uk-and-ireland/2018/02/23/actor-finds-document-teilhard-forced-sign-affirming-church-teaching/.

18  Marshall Savage, *The Millennial Project: Colonizing the Galaxy in Eight Easy Steps* (Boston: Little, Brown, and Company, 1992), 356.

19  Savage, *The Millennial Project*, 17.

20  Savage, *The Millennial Project*, 19.

21  Savage, *The Millennial Project*, 23.

22  Savage, *The Millennial Project*, 328.

23  Savage, *The Millennial Project*, 371.

24  Savage, *The Millennial Project*, 356.

25  Carl Sagan, *Pale Blue Dot: A Vision of the Human Future in Space* (New York: Random House, 1994), 264.

26  Dmitri Donskoy, "Space Environments Ecovillage" website, http://dmitridonskoy.com/see/faq.htm (Accessed April 18, 2019).

27  "6 Billion, the Game of the New Millennium," http://members.optusnet.com.au/bnbg6billion/6billion.htm.

28  Emily Pollock, "UN Brings Back Controversial Floating City Concept," *Engineering.com*, April, 15, 2019, https://www.engineering.com/BIM/ArticleID/18941/UN-Brings-Back-Controversial-Floating-City-Concept.aspx.

29  Savage, *The Millennial Project*, 357.

30  Savage, *The Millennial Project*, 361.

31  Savage, *The Millennial Project*, 355.

32   Br. Guy Consolagmo, email to author, April 1, 2019.

33   Rusty Schweickart, *L5-News* October, 1977, vol. 2, no. 10, 2.

34   Stephen Hawking and Leonard Mlodinow, *The Grand Design* (New York: Bantam Books, 2010), 153–54.

35   Hawking and Mlodinow, 159.

36   Fred Hoyle, "The Universe: Past and Present Reflections," *Annual Review of Astronomy and Astrophysics* 20, (1982): 4.

37   Hoyle, "The Universe," 16.

38   Kurt Vonnegut Jr., *Breakfast of Champions* (New York: Dial Press, 2002 [1973]), 3.

39   Paul C. W. Davies, "Life, Mind, and Culture as Fundamental Properties of the Universe," in Steven J. Dick and Mark L. Lupisella, eds., *Cosmos and Culture: Cultural Evolution in a Cosmic Context* (Washington, DC: NASA, 2009), 385.

40   Hawking and Mlodinow, *The Grand Design*, 136.

41   Freeman Dyson, interview with author, October 3, 2019.

42   Mariette Le Roux and Laurence Coustal, "After Death, Hawking Cuts 'Multiverse' Theory Down to Size," May 6, 2018, *Phys Org News*, https://phys.org/news/2018-05-death-hawking-multiverse-theory-size.html.

43   Le Roux and Coustal, "After Death, Hawking Cuts 'Multiverse'."

44   Dan Peterson, "The Vatican's Astronomer on God and the Stars," *Patheos*, December 22, 2018, https://www.patheos.com/blogs/danpeterson/2018/12/the-vaticans-astronomer-on-god-and-the-stars.html.

45   Interview with Br. Guy Consolagno on Connect 5, October 26, 2016, https://www.youtube.com/watch?v=AV3fPU7ZpEU.

46   Wendell Mendell, "Space Activism as an Epiphanic Belief System," in Steven J. Dick, ed., *Historical Studies in the Societal Impact of Spaceflight* (Washington DC: NASA, 2015), 581.

47   Sagan, *Pale Blue Dot*, 404.

48   Rabbi Steven Cohen of Congregation B'nai B'rith Congregation in Santa Barbara, California, helped me track down this quote.

49   Olaf Stapledon, "Interplanetary Man?" *Journal of the British Interplanetary Society* 7, no. 6 (1948), 22.

CHAPTER ELEVEN—THE SPACE RAVE

1   Nell Greenfield Boyce, "After 50 Years, Space Monkeys Not Forgotten," *NPR online*, May 28, 2009, https://www.npr.org/templates/story/story.php?storyId=104578202.

2   "Religion and Science," Pew Research Center, October 2015. This survey polled 2,002 adults in August 2014.

3   This result was gleaned a year after Yuri's Night, but probably accurate for 2018 as well. "Attitudes Toward Space Exploration," C-Span/Ipsos Poll, 2019, https://www.ipsos.com/en-us/news-polls/cspan-space-exploration-2019.

4   "Majority of Americans Believe It Is Essential that the U.S. Remain a Global Leader in Space," Pew Research Center, June, 2018.

# INDEX

$\bullet\ \bullet\ \bullet$